GRAFT-TRANSMISSIBLE DISEASES OF GRAPEVINES

Handbook for detection and diagnosis

Edited by
G.P. Martelli
Professor of Plant Virology
University of Bari, Italy

INTERNATIONAL COUNCIL FOR THE STUDY OF VIRUSES
AND VIRUS DISEASES OF THE GRAPEVINE

FOOD AND AGRICULTURE ORGANIZATION OF THE UNITED NATIONS
Rome, 1993

M-14

ISBN 92-5-103245-9

Foreword

This handbook is the second of a series. Like its companion volume, *Graft-transmissible diseases of citrus – Handbook on detection and diagnosis*, compiled by C.N. Roistacher, it aims to illustrate current methods for detecting and identifying viruses and virus-like diseases of grapevines (*Vitis vinifera* L., American *Vitis* species and their hybrids). Emphasis is placed on biological detection methods, i.e. the use of woody indicators and herbaceous differential hosts, which are still the basis for diagnosis and can readily be applied in countries with minimum laboratory facilities.

However, as many grapevine pathogens can also be detected and identified by serology, electron microscopy, gel electrophoresis and molecular hybridization, these techniques are described in detail in Part III.

Detailed further information on virus and virus-like diseases of the grapevine is available in a number of books and review articles, such as:

- **Bovey, R., Gartel, W., Hewitt, W.B., Martelli, G.P. & Vuittenez, A.** 1980. *Virus and virus-like diseases of grapevines.* Lausanne, Payot. 181 pp.
- **Bovey, R. & Martelli, G.P.** 1992. *Directory of major virus and virus-like diseases of grapevines.* Mediterranean Fruit Crop Improvement Council/ ICVG. 111 pp.
- **Pearson, R.G. & Goheen, A.C.** 1988. *Compendium of grape diseases.* St Paul, MN, USA, American Phytopathological Society Press. 93 pp.
- **Uyemoto, J., Martelli, G.P., Woodham, R.C., Goheen, A.C. & Dias, H.F.** 1978. Grapevine (*Vitis*) virus and virus-like diseases. *In* O.W. Barnett & S.A. Tolin, eds. *Plant virus slide series*, Set 1. Clemson, SC, USA, Clemson University. 100 slides, 29 pp.

In addition, readers are referred to bibliographic lists of papers on virus and virus-like diseases of grapevines to 1984 compiled under the auspices of the International Council for the Study of Viruses and Virus Diseases of the Grapevine (ICVG):

- **Bovey, R. & Martelli, G.P.** 1986. The viroses and virus-like diseases of the grapevine. A bibliographic report, 1979-1984. *Vitis*, 25: 227-275.
- **Caudwell, A.** 1965. *Bibliographie des viroses de la vigne des origines à 1965.* Paris, Office international de la vigne et du vin. 76 pp.
- **Caudwell, A., Hewitt, W.B. & Bovey, R.** 1972. Les viroses de la vigne. Bibliographie de 1965-1970. *Vitis*, 11: 303-324.
- **Hewitt, W.B. & Bovey, R.** 1979. The viroses and virus-like diseases of the grapevine. A bibliographic report, 1971-1978. *Vitis*, 18: 316-376.

Preface

Grapevine is widely cultivated in many countries of the world. The crop contributes to the well-being of farmers, workers, the industrial sector and the economy of countries. Demands made on the products and by-products of grapevine are constantly increasing in both local and foreign markets. This stimulates expansion of viticulture and encourages efforts toward the improvement of the quality and quantity of the produce. Such efforts necessitate the exchange of germplasm, the introduction of new varieties and the improvement of old varieties. These, if not carefully managed, could have a very negative effect upon the sanitary situation of the crop.

Virus and virus-like diseases constitute a major limiting factor to the development, quality and productivity of grapevine. Once a plant is infected, there is no cure. Therefore, the only way to secure a healthy crop is to ensure that the planting material is healthy (virus-tested) and that the factors that could contribute to reinfestation are eliminated.

Only a few institutions in the world have substantial programmes on the recovery and maintenance of healthy grapevine planting material and on detection and diagnosis of grapevine virus and virus-like diseases. Nevertheless, a wealth of relevant technology is now available. This ranges from simple methods requiring basic facilities (e.g. biological indexing) to sophisticated methods involving advanced equipment, chemicals and supplies and well-trained personnel for laboratory indexing. In both cases, the available technology could be easily transferred to those countries that intend to develop their viticulture through production, maintenance and distribution of healthy (virus-tested) planting material.

In order to make this technology universally available, the Food and Agriculture Organization of the United Nations (FAO) and the International Council for the Study of Viruses and Virus Diseases of the Grapevine (ICVG) have cooperated in the preparation and publication of this handbook, which reflects in a simple and direct way the practical knowledge required and available for detection and diagnosis of virus and virus-like diseases of grapevine.

The handbook was compiled by G.P. Martelli, Professor of Plant Virology, University of Bari, Italy. Professor Martelli has accumulated, over the last three decades, a vast wealth of experience concerning viticulture virology and viticulture phytosanitation schemes and has had a significant role in initiating and developing grapevine certification programmes in Italy and other European countries. The other contributors to the handbook are distinguished authorities on virology from various parts of the world who have developed or contributed to the techniques reported in this publication.

I wish to express my gratitude to Professor Martelli and the other contributing scientists for their willingness to share their knowledge and experience.

FAO is confident that this publication will fill an existing gap in the literature and will significantly contribute to the development of healthy vineyards.

N.A. Van der Graaff
Chief, FAO Plant Protection Service

Contents

Acknowledgements

I wish to thank all those who have kindly supplied information and slides: A. Brun, France; B. Di Terlizzi, G. Granata and U. Prota, Italy; H. Andris, J. Clark, M.K. Corbett, D. Gonsalves, P. Goodwin, M.G. Kinsey, A. Purcell, D.C. Ramsdell and A. Yen, USA; M. Rudel, Germany.

Great appreciation and thanks are also expressed to Dr M.M. Taher, FAO Regional Plant Protection Officer for the Near East, for his encouragement and many useful suggestions; to Dr L. Chiarappa, former Chief of the FAO Plant Protection Service, who promoted and fostered the realization of this handbook, devoting much time and effort to its layout; and to Dr W.B. Hewitt, Professor Emeritus, University of California, for revising and improving the manuscript.

G.P. Martelli
Editor

Contributors

A.D. Avgelis
Plant Protection Institute
Heraklion, Crete
Greece

D. Boscia
Centro di Studio del CNR sui Virus e le
Virosi delle Colture Mediterranee
Bari
Italy

M. Cambra
Instituto Valenciano de
Investigaciones Agrarias
Valencia
Spain

T. Candresse
Station de pathologie végétale
INRA
Villenave-D'Ornon
France

A. Caudwell
Station de recherches sur les
mycoplasmes et les arbovirus
des plantes
INRA
Dijon
France

J. Dunez
Station de pathologie végétale
INRA
Villenave-D'Ornon
France

S.M. Garnsey
USDA Hort. Research Laboratory
Orlando, FL
USA

D.A. Golino
Dept of Plant Pathology
University of California
Davis, CA
USA

R.F. Lee
Citrus Research & Education Center
Lake Alfred, FL
USA

J. Lehoczky
Plant Protection Institute
Budapest
Hungary

G. Macquaire
Station de pathologie végétale
INRA
Villenave-D'Ornon
France

S. Namba
Laboratory of Plant Pathology
University of Tokyo
Japan

C.N. Roistacher
Dept of Plant Pathology
University of California
Riverside, CA
USA

I.Ch. Rumbos
Plant Protection Institute
Volos
Greece

V. Savino
Dipartimento di Protezione delle Piante
dalle Malattie
University of Bari
Italy

J.S. Semancik
Dept of Plant Pathology
University of California
Riverside, CA
USA

B. Walter
Station de recherche vigne et vin
INRA
Colmar
France

Abbreviations

AILV
Artichoke Italian latent nepovirus

AGVd
Australian grapevine viroid

AMV
Alfalfa mosaic virus

APS
Ammonium persulphate

ArMV
Arabis mosaic nepovirus

BBLMV
Broad bean leaf mottle nepovirus

BBWV
Broad bean wilt fabavirus

BSA
Bovine serum albumen

CarMV
Carnation mottle carmovirus

cDNA
Complementary deoxyribonucleic acid

CEVd
Citrus exocortis viroid

CMV
Cucumber mosaic cucumovirus

DAS ELISA
Double antibody sandwich ELISA

DAS-I ELISA
Double antibody sandwich indirect ELISA

DIECA
Diethyldithiocarbamate sodium

DNA
Deoxyribonucleic acid

dPAGE
Denaturing polyacrylamide gel electrophoresis

dsRNA
Double-stranded ribonucleic acid

DTT
Dithiothreitol

EDTA
Ethylenedinitrilotetraacetate

ELISA
Enzyme-linked immunosorbent assay

EM
Electron microscopy

FD
Flavescence dorée

GALV
Grapevine Algerian latent tombusvirus

GAV
Grapevine ajinashika virus

GBLV
Grapevine Bulgarian latent nepovirus

GCMV
Grapevine chrome mosaic nepovirus

GFkV
Grapevine fleck virus

GFLV
Grapevine fanleaf nepovirus

GLPV
Grapevine line pattern virus

GLRaV I, II, III, IV, V
Grapevine leafroll-associated closteroviruses I, II, III, IV, V

GSV
Grapevine stunt virus

GTRV
Grapevine Tunisian ringspot nepovirus

GVA
Grapevine closterovirus A

GVB
Grapevine closterovirus B

GVd-c
Grapevine viroid – cucumber

GYSVd 1 and 2
Grapevine yellow speckle viroids 1 and 2

HSVd
Hop stunt viroid

IEM
Immunoelectron microscopy

IgG
Immunoglobulin G

IgM
Immunoglobulin M

ISEM
Immunosorbent electron microscopy

LR
Leafroll disease

MLO
Mycoplasma-like organism

OD
Optical density

PAGE
Polyacrylamide gel electrophoresis

PAMV
Petunia asteroid mosaic tombusvirus

PBS
Phosphate-buffered saline

PBST
Phosphate-buffered saline Tween 20

PD
Pierce's disease

PEG
Polyethylene glycol

PRMV
Peach rosette mosaic nepovirus

PVP
Polyvinylpyrrolidone

RNA
Ribonucleic acid

RRV
Raspberry ringspot nepovirus

RW
Rugose wood complex

SDS
Sodium dodecyl sulphate

SLRV
Strawberry latent ringspot nepovirus

SoMV
Sowbane mosaic sobemovirus

STE
Saline-tris-EDTA buffer

TBRV
Tomato black ring nepovirus

TBS
Tris-buffered saline

TEMED
Tetramethylethylenediamine

TMV
Tobacco mosaic tobamovirus

TNV
Tomato necrosis necrovirus

TomRSV
Tomato ringspot nepovirus

TRSV
Tobacco ringspot nepovirus

TSWV
Tomato spotted wilt bunyavirus

Introduction

Virus and virus-like diseases and other infectious diseases of grapevines (*Vitis* spp.) are induced by intracellular pathogens of various nature. These diseases are widespread throughout the world wherever grapevines, especially *Vitis vinifera*, are grown. Although their causal agents may be spread naturally by vectors (i.e. nematodes, pseudoccid mealybugs, leaf-hoppers), the major and most efficient means of dissemination of these diseases is through infected propagative material.

The following types of diseases are known:

TRUE VIRUS DISEASES

These diseases are induced by recognized viruses which have been isolated, identified and, in some cases, reinoculated into grapevines, reproducing the natural syndrome.

Some of the virus diseases known to date (Table 1) are caused by nepoviruses, of which two major groups can be recognized according to the geographical origin of the viruses and their nematode vectors. Notable exceptions are grapevine fanleaf virus (GFLV) and its major vector *Xiphinema index*, which are both probably native to ancient Asia Minor but now have a worldwide distribution because of unrestricted commercial trade.

Many additional viruses have been isolated by mechanical inoculation from grapevines (Table 2), and the list is growing steadily. Some of these viruses have no economic importance, but are occasional contaminants of vines grown in specific environments.

At least six different closteroviruses have been identified in vines affected by leafroll and/or the rugose wood complex. The frequency of these records leaves little doubt that closteroviruses are involved in the genesis of one or more such diseases. However, proof of cause or Koch's postulates must be fulfilled before any of these viruses can be identified as the aetiological agent of a specific disorder.

VIROID DISEASES

Viroids were first discovered in grapevines in 1984, but they are now known to have an extremely high incidence and a worldwide distribution. There are at least six viroids reported to infect grapevines in nature (Table 3). Three of these, i.e. hop stunt viroid (HSVd), citrus exocortis viroid (CEVd) and Australian grapevine viroid (AGVd), do not appear to cause diseases. Yellow speckle viroids are the only pathogenic viroids recognized so far. They may be implicated in the aetiology of the vein banding disease. Viroids are spread by grafting, by propagating material and by pruning tools.

VIRUS-LIKE DISEASES

Virus-like diseases are induced by unidentified agents that occur in the host tissues. They are perpetuated through propagative material and transmitted by grafting. As yet, no virus particles have been found associated with them. Some of these diseases are latent (e.g. vein necrosis) or semi-latent (e.g. vein mosaic and enations) in European grapes (*V. vinifera*) and most American *Vitis* species, so they can only be detected by graft inoculation to appropriate indicators. None has a recognized vector or is known to spread naturally.

TABLE 1

Virus and virus-like diseases of the grapevine

Disease	Geographical distribution
Major virus diseases	
Grapevine degeneration (fanleaf)	Worldwide
Grapevine degeneration (other European nepoviruses)	Europe, occasional records in Asia and Canada
Grapevine decline (American nepoviruses)	USA and Canada
Leafroll complex	Worldwide
Rugose wood complex (corky bark, rupestris stem pitting, Kober stem grooving, LN 33 stem grooving)	Worldwide
Minor virus diseases	
Yellow mottle (alfalfa mosaic virus)	Central and eastern Europe
Line pattern (grapevine line pattern virus)	Hungary
Yellow dwarf (tomato spotted wilt virus)	Taiwan
Stunt (isometric virus transmitted by the leafhopper *Arboridia apicalis*)	Japan
Ajinashika (phloem-limited, non-mechanically transmissible isometric virus)	Japan
Fleck (phloem-limited, non-mechanically transmissible isometric virus)	Worldwide
Roditis leaf discoloration	Greece
Viroid diseases	
Yellow speckle	Worldwide
Virus-like diseases	
Enations	Europe, America (USA, Venezuela), South Africa, New Zealand, Australia
Vein necrosis	Europe, Mediterranean basin, USA (California)
Vein mosaic/summer mottle	Europe, Australia
Asteroid mosaic	USA (California), Greece

DISEASES INDUCED BY PHLOEM- OR XYLEM-LIMITED PROKARYOTES

Two different groups of such diseases are known (Table 4), those caused by non-cultivable mollicutes formerly known as mycoplasma-like organisms (MLOs), i.e. flavescence dorée and other grapevine yellows not transmitted by the leafhopper *Scaphoideus titanus*; and those caused by intraxylematic bacteria such as *Xylella fastidiosa*, the agent of Pierce's disease.

TABLE 2

Geographical distribution and vectors of viruses known to infect grapevines

Virus	Geographical distribution	Vector
Mechanically transmissible viruses		
Artichoke Italian latent nepovirus (AILV)	Bulgaria	*Longidorus apulus, Longidorus fasciatus*
Alfalfa mosaic virus (AMV)	Central and Eastern Europe	Aphids
Arabis mosaic nepovirus (ArMV)	Europe (Switzerland, Germany, Hungary, Yugoslavia, Bulgaria, France, Italy), Japan	*Xiphinema diversicaudatum*
Broadbean wilt fabavirus (BBWV)	Bulgaria, South Africa	Aphids
Blueberry leaf mottle nepovirus (BBLMV)	USA (New York)	Unknown
Carnation mottle carmovirus (CarMV)	Greece	Unknown
Cucumber mosaic cucumovirus (CMV)	Denmark	Aphids
Grapevine Algerian latent tombusvirus (GALV)	Algeria	Unknown
Grapevine Bulgarian latent nepovirus (GBLV)	Bulgaria, Portugal, Yugoslavia	Unknown
Grapevine chrome mosaic nepovirus (GCMV)	Hungary, Yugoslavia	Unknown
Grapevine fanleaf nepovirus (GFLV)	Worldwide	*Xiphinema index, Xiphinema italiae*
Grapevine line pattern ilarvirus (GLPV)	Hungary	Unknown
Grapevine Tunisian ringspot nepovirus (GTRV)	Tunisia	Unknown
Grapevine closterovirus A (GVA)	Europe, Mediterranean	*Planococcus ficus, Planococcus citri, Pseudococcus longispinus*
Peach rosette mosaic nepovirus (PRMV)	USA (Michigan), Canada (Ontario)	*Xiphinema americanum, Longidorus diadecturus*
Petunia asteroid mosaic tombusvirus (PAMV)	Germany, Italy, Czechoslovakia	Unknown
Potato X potexvirus (PVX)	Italy, Tunisia	Unknown
Raspberry ringspot nepovirus (RRV)	Germany	*Longidorus macrosoma, Longidorus elongatus*
Strawberry latent ringspot nepovirus (SLRV)	Germany, Italy, Turkey	*Xiphinema diversicaudatum*
Sowbane mosaic sobemovirus (SoMV)	Germany, Czechoslovakia	Unknown
Tobacco mosaic tobamovirus (TMV)	Europe (Germany, Bulgaria, Italy, Yugoslavia, former Soviet Union), USA	Unknown
Tobacco ringspot nepovirus (TRSV)	USA (New York)	*Xiphinema americanum*

(continued)

TABLE 2 (cont.)

Virus	Geographical distribution	Vector
Tomato blackring nepovirus (TBRV)	Germany, Israel, Canada (Ontario)	*Longidorus attenuatus, Longidorus elongatus*
Tomato ringspot nepovirus (TomRSV)	USA (California and New York), Canada (Ontario)	*Xiphinema californicum, Xiphinema americanum, Xiphinema rivesi*
Tobacco necrosis necrovirus (TNV)	South Africa	*Olpidium brassicae*
Tomato spotted wilt virus (TSWV)	Taiwan Province (China)	Thrips
Non-mechanically transmissible viruses		
Grapevine fleck virus (GFkV)	Worldwide	Unknown
Grapevine ajinashika virus (GAV)	Japan	Unknown
Grapevine stunt virus (GSV)	Japan	*Arboridia apicalis*
Grapevine leafroll-associated closterovirus I	Europe, Mediterranean, USA	Unknown
Grapevine leafroll-associated closterovirus II	Europe, Mediterranean	Unknown
Grapevine leafroll-associated closterovirus III	Europe, Mediterranean, USA	*Planococcus ficus, Pseudococcus longispinus*
Grapevine leafroll-associated closterovirus IV	USA, Mediterranean	Unknown
Grapevine leafroll-associated closterovirus V	Mediterranean	Unknown

Note: Three additional closteroviruses have been found which, apparently, are serologically unrelated to all the above. Of these, one is reported to be associated with corky bark and the remaining two appear to be associated with leafroll.

TABLE 3
Viroids of grapevines

Viroid	Number of nucleotides	Geographical distribution
Hop stunt (HSVd)	297	Probably worldwide
Grapevine yellow speckle 1 (GYSVd-1)	367	Probably worldwide
Grapevine yellow speckle 2 (GYSVd-2)	363	Probably worldwide
Citrus exocortis A (CEVd)	371	Spain, Australia
Australian grapevine viroid (AGVd)	369	Australia
Grapevine viroid – cucumber (GVd-c)	not determined	USA (California)

TABLE 4

Grapevine diseases induced by intracellular prokaryotes (mycoplasma-like organisms and bacteria)

Disease	Agent	Vector	Geographical distribution
Flavescence dorée	MLO	*Scaphoideus titanus*	France, Italy
Grapevine yellows	Probably MLO	Unknown (probably leafhoppers)	Europe (France, Germany, Italy, Switzerland, Greece, Romania, Bulgaria), Chile, Israel, New Zealand, Australia
Pierce's disease	*Xylella fastidiosa* (bacterium)	Several species of leafhoppers	Central and North America

Part I

Detection and identification of specific grapevine diseases or pathogens

TRUE VIRUS DISEASES
Grapevine degeneration – fanleaf

G.P. Martelli

CAUSAL AGENT

The causal agent of fanleaf is grapevine fanleaf nepovirus (GFLV), a mechanically transmissible virus with isometric particles 30 nm in diameter and a bipartite genome (Quacquarelli *et al.*, 1976). GFLV populations are remarkably homogeneous serologically, so that natural serological variants are rarely found when screening is done by conventional serological techniques (Savino, Chérif and Martelli, 1985). Serological differences are detected, however, by monoclonal antibodies (Huss *et al.*, 1987). Biological variants differing according to the symptoms induced in artificially inoculated herbaceous hosts or in naturally infected grapevines are many.

GEOGRAPHICAL DISTRIBUTION

The disease and the causal virus have been recorded from all major grapevine-growing areas of the world. The level of infection is often very high, so that cultivars grown in certain viticultural districts are totally diseased.

ALTERNATE HOSTS

No alternate host is known. GFLV has never been found naturally in any wild or cultivated plant species other than *Vitis vinifera* and American *Vitis* species.

Described by Hewitt, 1954.

FIELD SYMPTOMS

Fanleaf degeneration is characterized by two distinct syndromes deriving from a differential reaction of the hosts to biologically distinct strains of the causal virus. Vein banding, another syndrome traditionally thought to be caused by chromogenic GFLV strains, may have a different aetiology, perhaps viroidal, as discussed under Viroid diseases.

Infectious malformations (fanleaf proper)

Distorting virus strains may cause vines to be stunted or less vigorous than normal. Leaves are variously and severely distorted, asymmetrical, cupped and puckered and exhibit acute dentations (Figures 1 and 2). Chlorotic mottling may sometimes accompany foliar deformations (Figure 3). Canes are also malformed, showing abnormal branching, double nodes, different length or exceedingly short internodes, fasciations and zigzag growth (Figures 4 and 5). Bunches are reduced in number and size, ripen irregularly and have shot berries and poor berry setting (Figure 6). Foliar symptoms develop early in the spring and persist through the vegetative season, although some masking may occur in summer.

Yellow mosaic

Yellow mosaic is caused by chromogenic virus strains. Affected vines show chrome-yellow discolorations that develop early in the spring and may affect all vegetative parts of the vines (leaves, herbaceous shoot axes, tendrils and inflorescences) (Figure 7). Chromatic alterations of the leaves vary from a few scattered yellow spots, sometimes appearing as rings and lines, to variously extended mottling of the veinal and/or interveinal areas, to total yellowing (Figures 8 to 11). In spring, affected plants in a vineyard can readily be spotted from a distance (Figure 12). Malformations of leaves and canes are usually not prominent, but clusters may be smaller than normal and may have shot berries. In hot climates the newly produced summer vegetation has a normal green colour, while the yellowing of the old growth turns whitish and tends to fade away (Figure 13).

Symptoms induced by distorting virus strains show up equally well under field and greenhouse conditions, whereas chrome-yellow alterations evoked by chromogenic strains may not develop in greenhouses.

NATURAL SPREAD

Spread at a site (i.e. within a vineyard or between adjacent vineyards) is mediated by nematodes. Two longidorid nematode vectors of GFLV are known: *Xiphinema index* and *Xiphinema italiae* (Hewitt, Raski and Goheen, 1958; Cohn, Tanne and Nitzany, 1970). The former species is by far the more efficient vector under natural and experimental conditions. Although not all *X. index* populations are equally efficient in transmitting virus isolates (Catalano, Roca and Castellano, 1989), this nematode is to be regarded as the major, if not the only, natural and economically important GFLV vector. It has a limited range of alternate natural hosts (e.g. fig, mulberry, rose), but these hosts are immune to GFLV. No natural virus reservoirs are known other than grapevines. GFLV persists in volunteer plants and in the roots of lifted vines that remain viable in the soil, constituting an important source of inoculum. Transmission of GFLV through grapevine seeds has been reported (Lazar, Kölber

and Lehoczky, 1990), but it has negligible epidemiological significance.

Long-distance spread is passively but efficiently effected through dissemination of infected propagating material (budwood and rooted cuttings of rootstocks) (Martelli, 1978).

DETECTION

Field symptoms are often obvious enough to allow the detection of diseased vines. Difficulties may be encountered, however, even by keen and experienced observers, in the presence of tolerant varieties and/or infections by mild virus strains. Observations for symptoms should be carried out twice during the vegetating season: in mid-spring for growth abnormalities, deformation and chromatic alterations of the foliage; and at leaf shedding for abnormalities of the canes.

IDENTIFICATION

Indexing by graft transmission

Although none of the *Vitis* species used as indicators is immune to GFLV, the quickest and most typical responses are given by *Vitis rupestris* St George. In greenhouse indexing (chip-budding, green grafting) at 22 to 24°C shock symptoms appear as soon as three to four weeks after grafting. The symptoms consist of chlorotic spots, rings and lines (Figure 14), sometimes accompanied by malformations and localized necrosis of the tissues. These symptoms are transient. They are followed by chronic reactions that, for distorting virus strains, consist of reduced growth and severely deformed leaves with prominent teeth (Figure 15). Chronic symptoms develop within two to three months from inoculation and persist throughout the vegetating season (Hewitt *et al.*, 1962; Martelli and Hewitt, 1963).

Field indexing (cleft or whip grafting) reactions are slower. Symptoms appear during the first year of vegetation. They are of the chronic type for distorting virus strains but consist of various patterns of yellow discolorations, sometimes accompanied by leaf deformity (Figure 16) for chromogenic virus strains. With certain virus sources yellowing may appear in the second year after inoculation. Yellow discolorations do not usually develop in greenhouse indexing.

Indicators other than *V. rupestris* may react to GFLV infections with chronic-type responses. These appear in the first or second year after grafting and consist of deformations and/or bright yellowing of the foliage. The type and severity of the symptoms vary with the nature and virulence of the virus strain.

Transmission to herbaceous hosts

GFLV is readily transmitted to herbaceous hosts by inoculation of expressed sap, provided that some precautions are observed. The best inoculum sources are young, tender symptomatic leaves of the spring flush collected directly in the field or from greenhouse-grown cuttings. Young, succulent roots are equally good, if not superior, sources of inoculum. Suitable root material is readily obtained by forcing grape cuttings to root in sand in a tray or bench heated to 24 to 25°C. Old leaves collected in the field in summer or autumn, especially in hot climates, constitute a poor source of virus, even if they show symptoms. Their use may result in repeated failures to transmit the virus.

For inoculation, 3 to 5 g of tissues (either leaves or roots) are ground in a chilled mortar together with an equal volume of one of the extraction media described in Part II. Media containing nicotine are recommended when the inoculum consists of leaves, especially if they are aged, whereas ordinary phosphate buffer is quite suitable for young root tips.

The herbaceous host range of GFLV is fairly wide, comprising some 50 species in seven dicotyledonous families (Dias, 1963; Hewitt *et*

al., 1962; Martelli and Hewitt, 1963; Taylor and Hewitt, 1964). Differential diagnostic hosts are:

- *Chenopodium amaranticolor* and *Chenopodium quinoa*, both reacting with occasional chlorotic/necrotic local lesions seven to ten days after inoculation, followed by systemic mottling and deformation of the leaves (Figures 17 and 18);
- *Gomphrena globosa*, which exhibits chlorotic local lesions about one week after inoculation, turning reddish with age, light green to yellow spots and twisting of systemically invaded opposite upper leaves (Figure 19);
- *Nicotiana benthamiana*, which reacts with occasional faint yellowish lesions followed by systemic mottling and deformation of the leaves in 10 to 15 days.

Whereas the above hosts are infected symptomatically by most virus isolates, many additional herbaceous indicators, such as *Phaseolus vulgaris*, *Cucumis sativus* and *Petunia hybrida*, are more selective in their susceptibility, being infectible by a smaller number of strains.

Presence of trabeculae (endocellular cordons)

Trabeculae are intraxylematic ribbon-like inclusions that cross tracheary elements radially. These structures are readily observed in cross-sectioned basal internodes of green or mature canes. Sections made by hand with a razor-blade are mounted without staining in a 1:1 mixture of water and glycerol and observed at low power under a light microscope. Freshly cross-sectioned canes can also be observed directly with a good (10 to 12x) magnifying lens. Trabeculae are more frequently found in American rootstocks than in European scion varieties. Their presence can be taken as evidence of viral infection, but their absence does not guarantee freedom from GFLV.

Serology

Immunodiffusion in agar gel is not applicable to leaf extracts of GFLV-infected grapevines, regardless of the severity of symptoms shown and the conditions under which plants are grown (field or greenhouse). Positive reactions are more readily and consistently obtained using crude sap expressed directly, without addition of buffers, from leaves of systemically invaded herbaceous hosts. Young upper leaves with strong symptoms are the best source of antigen.

ELISA. Immunoenzymatic procedures (see details in Part III) are now routinely employed for detection of GFLV in field- or greenhouse-grown vines (Bovey, Brugger and Gugerli, 1980; Engelbrecht, 1980; Walter *et al.*, 1984). Sources of antigens can be buds, roots, leaves and wood shavings, the last two being the most commonly used. Reactants can be either polyclonal antibodies or virus-specific or strain-specific monoclonal antibodies (Huss *et al.*, 1986, 1987).

The amount of plant material needed can be as low as 100 to 200 mg. However, it is customary to use 1 g of leaf tissues and 0.5 g of wood shavings. These are obtained by removing the bark and scratching, with a scalpel or a razor-blade, the cortex of mature canes freshly harvested or cold-stored at 4 to 6°C (Walter and Etienne, 1987). Tissue samples, regardless of whether taken from leaves or cortex, are ground in PBS-Tween containing 2 percent PVP and 1 to 2.5 percent nicotine. Addition of nicotine is considered of paramount importance if tests are to be made from leaf tissues. No nicotine is needed if the extraction medium is made up of 0.1 M Tris-HCl buffer, pH 8.2, containing 0.8 percent NaCl and 2 percent PVP (Walter and Etienne, 1987).

The advantage of wood shavings over leaves is twofold: they can be used throughout the year without the apparent loss of efficiency that arises because of seasonal variations of antigen titre in vegetating organs; and they give low and consistent background readings. If leaves are employed, it is important, if not mandatory, to use extracts of the same rootstock or European scion variety as negative and positive controls. There is, in fact, a marked difference in background readings between American rootstocks and European cultivars and, among *V. vinifera* varieties, between those with hairy and those with glabrous leaves.

Immune electron microscopy (IEM and ISEM). GFLV can be detected by ISEM following the procedure outlined in Part III. Consistent and satisfactory results are obtained in spring from upper symptomatic leaves (Bovey, Brugger and Gugerli, 1980; Russo, Martelli and Savino, 1980).

Molecular hybridization

Cloned cDNA probes have been prepared to genomic (RNA-1 and RNA-2) and satellite RNAs of GFLV. These probes were successfully used for identification of GFLV infections in field-grown grapevine leaf extracts denatured with formaldehyde (Fuchs *et al.*, 1991) and processed as described in Part III.

SANITATION

Sanitary selection and heat therapy together are powerful tools for reducing the incidence of fanleaf in newly established vineyards. Virus-free material (Figure 20) is readily obtained through conventional (Goheen and Luhn, 1973) or modified (Stellmach, 1980) heat therapy applications, micrografting and *in vitro* meristem tip or shoot-tip culture (Barlass *et al.*, 1982).

REFERENCES

Barlass, M., Skene, K.G.M., Woodham, R.C. & Krake, R.C. 1982. Regeneration of virus-free grapevines using *in vitro* apical culture. *Ann. Appl. Biol.*, 101: 291-295.

Bovey, R., Brugger, J.J. & Gugerli, P. 1980. Detection of fanleaf virus in grapevine tissue extracts by enzyme-linked immunosorbent assay (ELISA) and immune electron microscopy (IEM). *Proc. 7th Meet. ICVG*, Niagara Falls, NY, USA, 1980, p. 259-275.

Catalano, L., Roca, F. & Castellano, M.A. 1989. Efficiency of transmission of an isolate of grapevine fanleaf virus (GFV) by three populations of *Xiphinema index* (Nematoda: Dorylaimidae). *Nematol. Mediterr.*, 17: 13-15.

Cohn, E., Tanne, E. & Nitzany, F. 1970. *Xiphinema italiae*: a new vector of grape fanleaf virus. *Phytopathology*, 60: 181-182.

Dias, H.F. 1963. Host range and properties of grapevine fanleaf and grapevine yellow mosaic viruses. *Ann. Appl. Biol.*, 51: 85-95.

Engelbrecht, D.J. 1980. Indexing grapevines for grapevine fanleaf virus by enzyme-linked immunosorbent assay. *Proc. 7th Meet. ICVG,* Niagara Falls, NY, USA, 1980, p. 277-282.

Fuchs, M., Pinck, M., Etienne, L., Pinck, L. & Walter, B. 1991. Characterization and detection of grapevine fanleaf virus using cDNA probes. *Phytopathology*, 81: 559-565.

Goheen, A.C. & Luhn, C.F. 1973. Heat inactivation of viruses of grapevines. *Riv. Patol. Veg. Sci.*, 9: 287-289.

Hewitt, W.B. 1954. Some virus and virus-like diseases of grapevines. *Bull. Calif. Dept. Agric.*, 43: 47-64.

Hewitt, W.B., Goheen, A.C., Raski, D.J. & Gooding, G.V. Jr. 1962. Studies on virus diseases of the grapevine in California. *Vitis*, 3: 57-83.

Hewitt, W.B., Raski, D.J. & Goheen, A.C. 1958. Nematode vector of soil-borne fanleaf virus of grapevines. *Phytopathology*, 48: 586-595.

Huss, B., Muller, S., Sommermeyer, G., Walter, B. & Van Regenmortel, M.H.V. 1987. Grapevine fanleaf virus monoclonal antibodies: their use to distinguish different isolates. *J. Phytopathol.*, 119: 358-370.

Huss, B., Walter, B., Etienne, L. & Van Regenmortel, M.H.V. 1986. Grapevine fanleaf virus detection in various grapevine organs using polyclonal and monoclonal antibodies. *Vitis*, 25: 178-188.

Lazar, J. Kölber, M. & Lehoczky, J. 1990. Detection of some nepoviruses (GFV, GFV-YM, GCMV, ArMV) in the seeds and seedlings of grapevine by ELISA. *Kertgasdasag*, 22(4): 58-72.

Martelli, G.P. 1978. Nematode-borne viruses of grapevine, their epidemiology and control. *Nematol. Mediterr.*, 6: 1-27.

Martelli, G.P. & Hewitt, W.B. 1963. Comparative studies on some Italian and Californian virus diseases of grapevine. *Phytopathol. Mediterr.*, 2: 275-284.

Quacquarelli, A., Gallitelli, D., Savino, V. & Martelli, G.P. 1976. Properties of grapevine fanleaf virus. *J. Gen. Virol.*, 32: 349-360.

Russo, M., Martelli, G.P. & Savino, V. 1980. Immunosorbent electron microscopy for detecting sap-transmissible viruses of grapevine. *Proc. 7th Meet. ICVG*, Niagara Falls, NY, USA, 1980, p. 251-257.

Savino, V., Chérif C. & Martelli, G.P. 1985. A natural serological variant of grapevine fanleaf virus. *Phytopathol. Mediterr.*, 24: 29-34.

Stellmach, G. 1980. Moderate heat propagation of grapevines for eliminating graft transmissible disorders. *Proc. 7th Meet. ICVG,* Niagara Falls, NY, USA, 1980, p. 325-328.

Taylor, R.H. & Hewitt, W.B. 1964. Properties and serological relationships of Australian and

Californian soil-borne viruses of the grapevine, and arabis mosaic virus. *Aust. J. Agric. Res.*, 15: 571-585.

Walter, B. & Etienne, L. 1987. Detection of the grapevine fanleaf virus away from the period of vegetation. *J. Phytopathol.*, 120: 355-364.

Walter, B., Vuittenez, A., Kuszala, J., Stocky, G., Burckard, J. & Van Regenmortel, M.H.V. 1984. Détection sérologique du virus du court-noué de la vigne par le test ELISA. *Agronomie*, 4: 527-534.

Summary: fanleaf detection

GRAFT TRANSMISSION
Indicator
Vitis rupestris St George
No. plants/test
3-5 rooted cuttings
Inoculum
Wood chips, single buds, bud sticks, shoot tips
Temperature
22-24°C
Symptoms
Acute phase (shock) symptoms. Chlorotic spots, rings and lines, localized necrosis 3-4 weeks after grafting (chip-bud or green grafting);
Chronic symptoms. Reduced growth, severely deformed leaves with prominent teeth (distorting strains), yellow discolorations and mild deformation of the leaves (chromogenic strains)

TRANSMISSION TO HERBACEOUS HOSTS
Diagnostic host
Gomphrena globosa
Inoculum
Tissues from young symptomatic leaves or succulent roots
Extraction
Grind in 2.5 percent aqueous nicotine
Temperature
Below 25°C
Symptoms
Chlorotic local lesions soon turning reddish in 7-8 days; twisting of the upper leaves in 10-12 days

OTHER TESTS
Serology (ELISA, ISEM)
Molecular hybridization

FIGURE 1
Severely malformed leaves and bushy vegetation in a vine infected by a distorting GFLV strain

FIGURE 4
Abnormal branching in a shoot of a fanleaf-infected vine

FIGURE 2
Grape leaf with typical fanleaf symptoms

FIGURE 5
Abnormal shoot in a fanleaf-infected vine

FIGURE 3
Chlorotic mottle and deformation caused by fanleaf infection

FIGURE 6
Bunches from a healthy (left) and a fanleaf-infected (right) vine. Note reduced size and extensive shot berry condition

FIGURE 7
Total yellowing of a spring shoot of a vine infected by a
chromogenic GFLV strain

FIGURE 9
Yellow mosaic symptoms in a leaf
of an American rootstock

FIGURE 10
Spring symptoms of yellow mosaic

FIGURE 8
Yellow rings and line patterns in a leaf infected
by a chromogenic GFLV strain
(Photo: B. Walter)

FIGURE 11
Totally yellow and stunted vine affected by yellow mosaic
next to a normally growing plant

FIGURE 12
A patch of vines with yellow mosaic seen from a distance

FIGURE 14
Shock symptoms (chlorotic rings and lines) in a
V. rupestris indicator graft-inoculated with
a fanleaf source

FIGURE 13
Yellow mosaic symptoms fading away
in late summer

FIGURE 15
Chronic symptoms (severely deformed leaves with
prominent teeth) in a *V. rupestris* indicator graft-inoculated
with a fanleaf source

FIGURE 16
Chrome yellow discolorations and deformations of leaves
of a *V. rupestris* indicator graft-inoculated with a
yellow mosaic source

FIGURE 17
Systemic mottling induced by GFLV in *Chenopodium amaranticolor*

FIGURE 19
Twisting of the upper leaves typically induced by GFLV infections in *Gomphrena globosa*

FIGURE 18
Systemic mottling and deformation of the leaves of *Chenopodium quinoa* infected by GFLV

FIGURE 20
Bunches from a fanleaf-infected vine before (left) and after (right) heat therapy

TRUE VIRUS DISEASES
Grapevine degeneration – European nepoviruses

G.P. Martelli and B. Walter

CAUSAL AGENTS

In addition to GFLV there are seven distinct nepoviruses: tomato black ring (TBRV), arabis mosaic (ArMV), raspberry ringspot (RRV), strawberry latent ringspot (SLRV), grapevine chrome mosaic (GCMV), grapevine Bulgarian latent (GBLV) and artichoke Italian latent (AILV) viruses, involved to a varying extent in the aetiology of this disease. All these viruses have isometric particles about 30 nm in diameter and the bipartite genome typical of the nepovirus group (Harrison and Murant, 1977). All possess natural serological variants, but in most cases strains infecting grapevines belong to a single serotype. Some of these viruses are serologically distantly related to one another or to other nepoviruses, i.e. GCMV to TBRV, ArMV to GFLV, GBLV to blueberry leaf mottle virus (reviewed by Martelli and Taylor, 1989). There are biological variants that elicit different symptoms in naturally and artificially infected hosts.

GEOGRAPHICAL DISTRIBUTION

European nepoviruses prevail in central and eastern Europe, where they occur in many natural and agricultural environments together with their vectors. In grapevines, ArMV is common in

certain areas of France (Charentes, Alsace) and Germany (Palatinate) and is more rarely found in Switzerland, Italy, Yugoslavia, Hungary and Bulgaria. TBRV, SLRV and RRV are largely restricted to Germany (Palatinate and Moselle), even though there are occasional records from other European (Italy, France, Portugal, Turkey) and non-European (Israel) countries. GCMV and GBLV prevail in eastern Europe (Yugoslavia, Czechoslovakia, former USSR, Hungary, Bulgaria).

ALTERNATE HOSTS

ArMV, TBRV, RRV, SLRV and AILV have a wide range of wild and cultivated alternate hosts that constitute their natural reservoirs. These hosts include annual and perennial weeds, vegetable crops, shrubs, ornamentals and fruit trees. No alternate hosts for GCMV and GBLV are known.

FIELD SYMPTOMS

Like GFLV, some European nepoviruses (e.g. ArMV and GCMV) have distorting and chromogenic strains that elicit different symptomatologies, characterized by chlorotic mottling, leaf and cane deformation or chrome-yellow discolorations of the foliage (Figures 21 to 25). The symptoms shown in the field by vines affected by these viruses are very similar to, if not indistinguishable from, those of GFLV-induced degeneration.

Described by Bercks and Stellmach, 1966; Martelli *et al.*, 1977; Martelli, Lehoczky and Quacquarelli, 1966.

A distinctive trait of the yellowing caused by GCMV is that, unlike GFLV yellowing, it shows up equally well in the field and greenhouse. Furthermore, vines infected by GCMV lack vigour, bear little or no crop and tend to decline and die within a few years after infection. Heavy yield losses (up to 80 percent of the crop) are associated with infections by other European nepoviruses such as SLRV, TBRV and RRV (Rudel, 1985). These viruses induce leaf deformity and yellowing (Figures 26 and 27). In the cultivars Gruner Sylvaner and Riesling, SLRV elicits a most peculiar reddish discoloration of the tip of the spring shoots (Figure 28). This discoloration disappears in summer, when the vines exhibit a more or less normal appearance but are virtually fruitless (Figure 29).

NATURAL SPREAD

Experimental evidence for nematode transmission from grape to grape has been obtained only for ArMV/*Xiphinema diversicaudatum* and TBRV/*Longidorus attenuatus*. The ecology of these viruses in vineyards follows the same pattern known for other crops, i.e. seed transmission in weeds and persistence of the pathogens in their seedlings, which constitute sources of inoculum and food for the vector (reviewed by Martelli, 1978).

Although there is plenty of visual evidence that RRV and GCMV spread naturally in vineyards in a manner suggestive of soil transmission, their vectors have not been identified. Long-distance spread is through infected propagative material (budwood, rootstock rootings, grafted vines).

Rooted plants are high-risk materials because they can carry viruses and vectors together and thus establish active infection foci in any new environment favourable to active multiplication of the nematode vectors.

DETECTION

The presence of the disease in vineyards is disclosed by its symptoms. Plants can be inspected for symptoms throughout the vegetating season, but special attention should be paid to the spring growth, which ordinarily expresses the strongest reactions. Infections by mild virus strains or tolerance in certain cultivars can result in attenuated forms of disease that may escape observation. In the field it is virtually impossible to establish whether any given symptomatology is induced by one virus or another, or by a mixed infection of two or more viruses.

Sorting and identification of viruses can only be done by biological assays and/or laboratory tests.

IDENTIFICATION

Indexing by graft transmission

Any of the current graft inoculation procedures described in Part II can be used.

Graft transmission of GCMV to *Vitis* species yields foliar distortions (Figure 30) and yellow discolorations (Figure 31) like those induced by other nepoviruses (GFLV in particular). Somewhat more specific reactions are given by the cultivars Pinot noir and Jubileum 75, both of which show extremely severe stunting and necrosis of the apex (Figure 32). No symptoms are produced in *Vitis rupestris* St George, which differentiates GCMV from GFLV. In field indexing, symptoms develop on the second year's growth (Lehoczky, 1985).

Siegfriedrebe (FS4 201-39) is regarded in Germany as the best indicator for ArMV, RRV and TBRV. Cane deformations and foliar discolorations appear a few weeks after inoculation, but these symptoms are not distinctive enough to separate viruses. The Pinot noir cultivar can also be used as an indicator for TBRV.

Transmission to herbaceous hosts

All European nepoviruses are readily transmitted to herbaceous hosts by sap inoculation. Sources and preparation of inoculum and inoculation procedures are the same as described for GFLV.

The separation of viruses based on differential host range responses is not easy and may pose problems even to experienced workers. Certain hosts, however, give reactions that provide useful hints for identification.

Symptoms in *Chenopodium quinoa* are generally milder for ArMV than for GFLV. They consist of systemic mottling and mild deformations of the leaves, with rare local lesions (Figure 33). ArMV can also induce striking chlorotic ringspotting in *Nicotiana glutinosa* (Figure 34).

TBRV is very severe in *C. quinoa*, producing distinct chlorotic/necrotic local lesions followed by systemic spread and necrosis of the plant tips (Figure 35). RRV induces local chlorotic or necrotic spots in *Nicotiana rustica*, followed by chlorotic/necrotic systemic rings, spots and line patterns.

SLRV may elicit chlorotic local lesions in inoculated *Cucumis sativus* cotyledons, followed by systemic interveinal chlorosis.

GCMV has fairly narrow host range (Martelli and Quacquarelli, 1972). Diagnostic hosts are:

- *Chenopodium quinoa*. Small chlorotic local lesions, sometimes with necrotic centre, severe systemic chlorosis, necrotic speckling and apical necrosis (Figures 36 and 37);
- *Datura stramonium*. No visible localized infections. Transient systemic, yellowish zonate spots;
- *Gomphrena globosa*. Chlorotic or reddish local lesions. Transient systemic mosaic and vein clearing;
- *Phaseolus vulgaris*. Systemic symptoms only; mosaic, chlorotic rings and specks, sometimes necrotic blotches.

GBLV also has a narrow host range (Martelli *et al.*, 1977). Diagnostic hosts are:

- *Chenopodium quinoa*. Chlorotic or necrotic local lesions in three to four days, followed by systemic leaf mottle and necrotic flecks (Figure 38);
- *Gomphrena globosa*. Reddish local lesions and deformation but not twisting of the upper leaves;
- *Nicotiana clevelandii*. Necrotic local lesions, systemic leaf mottling and stunting.

Serology

Immunodiffusion tests may be used successfully with extracts from GBLV-infected grapevine leaves, especially when collected from greenhouse-grown cuttings. The same does not apply to tissue extracts of vines infected by any other European nepovirus. However, with these other viruses positive reactions are obtained using crude sap expressed directly, without addition of buffered solutions, from leaves of systemically invaded herbaceous hosts. Young, upper symptomatic leaves are the best source of antigen.

ELISA. Immunoenzymatic procedures carried out as described in Part III can be employed for the detection and identification of different European nepoviruses in naturally infected vines (Tanne, 1980; Stellmach, 1985; Walter *et al.*, 1984; Kölber *et al.*, 1985). Sources of antigen can be buds, roots, leaves, bark and wood scrapings. The weight of tissues used varies from 200 mg to 1 g, and the extraction buffer, as discussed for GFLV, may or may not contain nicotine (1 to 2.5 percent). As with GFLV, the use of buds from dormant canes or wood shavings may bypass the problem posed by the seasonal variation of virus titre, which limits the utilization of ELISA to certain growth periods (Rudel, Alebrand and Altmayer, 1983; Kölber *et al.*,

1985). Single as well as mixed infections can be picked up by ELISA, and the viruses can be identified. Precautions for positive and negative controls are the same as with GFLV.

Immune electron microscopy (ISEM). Detection and identification of European nepoviruses by ISEM can be carried out, as outlined in Part III, using upper symptomatic leaves collected in spring (Russo, Martelli and Savino, 1980).

Molecular hybridization

A probe to ArMV has been prepared (Steinkellner *et al.*, 1989). cDNA probes to GFLV RNA-1 or RNA-2 can also detect some ArMV isolates.

SANITATION

The same techniques used for GFLV are applicable.

REFERENCES

Bercks, R. & Stellmach, G. 1966. Nachweis verschiedener Viren in Reisigkrankheiten Reben. *Phytopathol. Z.*, 65: 288-296.

Harrison, B.D. & Murant, A.F. 1977. *Nepovirus group*. Descriptions of Plant Viruses, No. 185. Kew, UK, Commonw. Mycol. Inst./Assoc. Appl. Biol.

Kölber, M., Beczner, L., Pacsa, S. & Lehoczky, J. 1985. Detection of grapevine chrome mosaic virus in field-grown vines by ELISA. *Phytopathol. Mediterr.*, 24: 135-140.

Lehoczky, J. 1985. Detection of grapevine chrome mosaic virus in naturally infected vines by indexing. *Phytopathol. Mediterr.*, 24: 129-134.

Martelli, G.P. 1978. Nematode-borne viruses of grapevine, their epidemiology and control. *Nematol. Mediterr.*, 6: 1-27.

Martelli, G.P., Gallitelli, A., Abracheva, P., Savino, V. & Quacquarelli, A. 1977. Some properties of grapevine Bulgarian latent virus. *Ann. Appl. Biol.*, 85: 51-58.

Martelli, G.P., Lehoczky, J. & Quacquarelli, A. 1966. Host range and properties of a virus associated with Hungarian grapevines showing macroscopic symptoms of fanleaf and yellow mosaic. *Proc. Int. Conf. Virus Vector Perennial Hosts and Vitis*, 1965, p. 389-401. Div. Agric. Sci., Univ. Calif., Davis.

Martelli, G.P. & Quacquarelli, A. 1972. *Grapevine chrome mosaic virus*. Descriptions of Plant Viruses, No. 103. Kew, UK, Commonw. Mycol. Inst./Assoc. Appl. Biol.

Martelli, G.P. & Taylor, C.E. 1989. Distribution of viruses and their nematode vectors. *Adv. Dis. Vector Res.*, 6: 151-189.

Rudel, M. 1985. Grapevine damage induced by particular virus-vector combinations. *Phytopathol. Mediterr.*, 24: 183-185.

Rudel, M., Alebrand, M. & Altmayer, B. 1983. Untersuchungen über den Einsatz des ELISA-Test zum Nachweis verschiedener Rebenviren. *Vein Wiss.*, 38: 177-185.

Russo, M., Martelli, G.P. & Savino, V. 1980. Immunosorbent electron microscopy for detecting sap-transmissible viruses of grapevine. *Proc. 7th Meet. ICGV*, Niagara Falls, NY, USA, 1980, p. 251-257.

Steinkellner, H., Himmler, G., Laimer, M., Mattenovich, D., Bisztroy, G. & Katinger, H. 1989. Konstruktion von cDNA von Arabis Mosaik Virus und deren Anwendung für Diagnose. *Mitt. Klosterneuburg Rebe Wein Obstbau Frücht.*, 39: 242-246.

Stellmach, G. 1985. ELISA testing of grapevine rootings reared from nepovirus-infected mother plants forced to rapid growth. *Phytopathol. Mediterr.*, 24: 123-124.

Tanne, E. 1980. The use of ELISA for the detection of some nepoviruses in grapevines. *Proc. 7th Meet. ICVG*, Niagara Falls, NY, USA, 1980, p. 293-296.

Walter, B., Vuittenez, A., Kuszala, A., Stocky, G., Burckard, J. & Van Regenmortel, M.H.V. 1984. Détection sérologique du virus du court-noué de la vigne par le test ELISA. *Agronomie*, 4: 527-534.

Summary: grapevine degeneration detection

GRAFT TRANSMISSION
Indicators
Several *Vitis vinifera* cultivars: Pinot noir, Jubileum 75 (GCMV); Siegfriedrebe (ArMV, RRV, TBRV)

No. plants/test
3-5 rooted cuttings
Inoculum
Wood chips, single buds, bud sticks, shoot tips
Temperature
22-24°C
Symptoms
Severe stunting and necrosis of the apex of Pinot noir in the second year of vegetation (GCMV); foliar discolorations and cane deformations of Siegfriedrebe within the first year after inoculation

TRANSMISSION TO HERBACEOUS HOSTS
Diagnostic hosts
Datura stramonium (GCMV), *Chenopodium quinoa* (TBRV, GBLV), *Nicotiana clevelandii* (RRV), *Cucumis sativus* (SLRV), *Nicotiana glutinosa* (ArMV)
Inoculum
Tissue from young symptomatic leaves
Extraction
Grind in 2.5 percent aqueous nicotine
Temperature
Below 25°C
Symptoms
In *D. stramonium* (CGMV), transient, systemic, yellowish zonate spots;
In *C. quinoa* (TBRV), necrotic local lesions in 6-8 days, followed by mosaic and necrosis of the plant tip in about 2 weeks;
In *C. quinoa* (GBLV), necrotic local lesions in 3-4 days, systemic chlorotic mottle and necrosis;
In *N. clevelandii* (RRV), necrotic local spots and rings in 5-7 days and systemic veinal necrosis;
In *C. sativus* (SLRV), chlorotic local lesions in 5-7 days and systemic interveinal chlorosis or necrosis in 10-12 days;
In *N. glutinosa* (ArMV), chlorotic ringspots

OTHER TESTS
Serology (ELISA, ISEM)
Molecular hybridization

FIGURE 21
Mottling and deformation of a grape leaf induced by a
distorting strain of ArMV
(Photo: M. Rudel)

FIGURE 24
Typical yellow discoloration caused by a chromogenic
strain of GCMV

FIGURE 22
Yellow mottling induced by a
chromogenic strain of ArMV

FIGURE 25
A row of vines with severe chrome mosaic symptoms

FIGURE 23
Chlorotic mottling and leaf deformation
induced by a distorting strain of GCMV

FIGURE 26
Yellowing and marginal necrosis in vine
infected by RRV
(Photo: M. Rudel)

FIGURE 28
Reddish discolorations of the shoot apex
typically induced by SLRV in spring in certain
cultivars (left) compared to healthy shoot (right)
(Photo: M. Rudel)

FIGURE 27
Leaf deformity and yellowing in a vine
infected by TBRV
(Photo: M. Rudel)

FIGURE 29
Summer symptoms of SLRV. The vegetation is apparently
normal but there is no crop
(Photo: M. Rudel)

FIGURE 30
Stunting, mottling and leaf deformation in a Siegfriedrebe
indicator graft-inoculated with GCMV
(Photo: J. Lehoczky)

FIGURE 32
Extremely severe stunting caused by GCMV infections
in the indicator Pinot noir two years after grafting
(Photo: J. Lehoczky)

FIGURE 31
Chrome mosaic symptoms in the indicator
Kober 5BB
(Photo: J. Lehoczky)

FIGURE 33
Mosaic mottle induced by ArMV in *C. quinoa*

FIGURE 36
Chlorotic local lesions induced by
GCMV in *C. quinoa*

FIGURE 34
Yellow spots and rings induced by ArMV in *N. glutinosa*

FIGURE 37
Severe systemic yellowing induced by GCMV
in *C. quinoa*

FIGURE 35
Necrotic local lesions and top
necrosis induced by TBRV in
C. quinoa

FIGURE 38
Chlorotic local lesions induced by SLRV
in *C. quinoa*

TRUE VIRUS DISEASES

Grapevine decline – American nepoviruses

G.P. Martelli

CAUSAL AGENTS

Four distinct nepoviruses, tomato ringspot (ToRSV), tobacco ringspot (TRSV), peach rosette mosaic (PRMV) and, to a much lesser extent, blueberry leaf mottle (BBLMV), are implicated in the genesis of this complex disease. These viruses have isometric particles about 30 nm in diameter and the bipartite genome typical of the nepovirus group (Harrison and Murant, 1977; Stace-Smith and Ramsdell, 1987; Martelli and Taylor, 1989).

All of these viruses have natural serological variants, but only ToRSV has two serologically distinguishable strains (Piazzolla *et al.*, 1985) that infect grapevines and induce different diseases. There are biological variants that elicit different symptoms in naturally and artificially inoculated hosts.

GEOGRAPHICAL DISTRIBUTION

Grapevine decline is typical in the northeastern United States (Michigan, New York, Maryland and Pennsylvania) and Canada (Ontario). ToRSV also occurs in grapes in a restricted area of central California.

ALTERNATE HOSTS

All four viruses have a wide range of wild and cultivated alternate hosts. These are both annual (weeds, vegetable crops) and perennial (weeds,

shrubs, ornamentals, fruit trees) plants that serve as natural virus reservoirs.

FIELD SYMPTOMS

Symptomatological responses of grapevines vary according to the species (i.e. *Vitis vinifera, Vitis labrusca,* interspecific hybrids), the infecting virus and the climatic conditions. ToRSV-induced decline affects European cultivars (especially if self-rooted) more severely in colder than in warmer climates. Newly infected vines have normal growth but show occasional chlorotic mottling and rings on the leaves of a few shoots (Figure 39). In the following years affected vines decline rapidly, showing small mottled and distorted leaves, short internodes, stunted growth, little fruit set with straggling and shelled clusters, or no yield at all (compare Figures 40 and 41). Death of the vines may ensue, favoured by winter injury.

In Maryland and California ToRSV affects the yield rather than the vine's growth, which is more vigorous than normal. Yield is affected either by formation of small clusters (e.g. little grape disease of cultivars Vidal blanc and Carignan) or by severe reduction in fruit setting, as in the case of yellow vein disease in California. Although no visible foliar symptoms are associated with little grape disease, yellow vein is characterized by chrome-yellow flecking along the veins extending to the interveinal tissues (Figure 42).

The symptoms of TRSV-induced decline, which only occurs in New York State and

Described by Gilmer, Uyemoto and Kelts, 1970; Gilmer and Uyemoto, 1972; Gooding and Hewitt, 1962; Ramsdell and Meyers, 1974; Uyemoto, Taschenberg and Hummer, 1977.

Pennsylvania, are the same as those of ToRSV in native cultivars, but in *V. vinifera* they are similar to the symptoms induced by European nepoviruses (Figures 43 and 44).

In Michigan, *V. labrusca* and French hybrids affected by PRMV show a progressive decline over several years, accompanied by leaf and cane deformations, straggling and shot-berried clusters (Figures 45 and 46).

NATURAL SPREAD

Except for BBLMV, whose means of natural spread is unknown, the causal agents of grapevine decline are transmitted through soils by longidorid nematodes. The vector situation of these viruses is not as clear-cut as that of GFLV and other European nepoviruses. It is summarized in Table 5 (reviewed by Martelli and Taylor, 1989).

The efficiency of transmission is high for all virus/vector combinations except for PRMV/ *Longidorus elongatus*. Natural virus reservoirs are primarily perennial weeds in which the viruses are endemic prior to the establishment of the vineyards; these weeds are also hosts for the vectors. All viruses are transmitted through seeds of weeds and grapevines, whose seedlings perpetuate the inoculum. Long-distance spread may take place through budwood from mildly infected vines, or from plants grown in warmer climates where vegetation is not much affected.

DETECTION

Detection is based on observation of field symptoms. Inspections for the identification of newly infected vines, which may show mild and localized responses, need to be carried out with special care. Summer, when bunches have reached full size, is the best time for detecting little grape and yellow vein diseases. Yellow vein can be confused with syndromes induced by chromogenic GFLV strains and vein banding

TABLE 5
Vectors of the causal agents of grapevine decline

Virus	Vector
Tomato ringspot virus Type strain (decline)	*Xiphinema americanum sensu stricto*, *Xiphinema rivesi*
Californian strain (yellow vein)	*Xiphinema californicum*
Tobacco ringspot virus	*Xiphinema americanum sensu lato*
Peach rosette mosaic virus	*Xiphinema americanum sensu stricto*, *Longidorus diadecturus*, *Longidorus elongatus*

disease. Differential traits are that yellow vein symptoms develop earlier than vein banding and are usually shown by fewer leaves (Goheen and Hewitt, 1962).

IDENTIFICATION
Indexing by graft transmission

Indicators for yellow vein are the European varieties Grenache and Carignan, which reproduce the field syndrome following chip-bud grafting. *Vitis rupestris* St George, Kober 5BB and cv. Mission, which do not show symptoms when inoculated with yellow vein sources, react symptomatically when grafted with chromogenic GFLV and thus serve as differential indicators for the two viruses.

Baco and Kober 5BB, when inoculated with ToRSV sources, react with shock symptoms (chlorotic blotching and rings) that appear a few weeks after grafting (Figures 47 and 48). These are followed by systemic chronic symptoms similar to those seen in the field. Chardonnay reproduces more or less the field symptomatology when inoculated with TRSV sources (Figure 49).

Transmission to herbaceous hosts

All American nepoviruses are readily transmitted to herbaceous hosts by inoculation of sap from young leaves or roots, expressed in the presence of 2.5 percent aqueous nicotine or phosphate-buffered solutions. Inoculation procedures and precautions are the same as outlined for European nepoviruses. In addition, the herbaceous hosts used for isolation and diagnosis are the same as those used for European nepoviruses. ToRSV and TRSV have similar experimental hosts, which react with comparable, if not identical, symptoms. Thus, for example, either virus may induce tip necrosis in *Phaseolus vulgaris* (Figure 50), chlorotic local lesions and systemic mottling in *Cucumis sativus* (Figure 51), chlorotic/necrotic local lesions in *Chenopodium amaranticolor* (Figure 52) and *Chenopodium quinoa* followed by systemic mottling (Figure 53), and necrotic local lesions in *Vigna unguicolata* (Figure 54). Both viruses also cause chlorotic/necrotic rings and line patterns in various *Nicotiana* species, which after four to six weeks enter a recovery phase in which newly produced leaves are symptomless but contain virus.

Diagnostic hosts for PRMV are:
- *C. amaranticolor* and *C. quinoa*. Chlorotic necrotic local lesions, systemic mottling, leaf deformation and tip necrosis (Figure 55);
- *Nicotiana tabacum*. Chlorotic local lesions and/or necrotic ringspots; systemic chlorotic ringspots. Symptomless infection may occur with some virus isolates.

Diagnostic hosts for BBLMV are:
- *C. quinoa*. Chlorotic local lesions, systemic mottle and tip necrosis;
- *Nicotiana clevelandii*. Necrotic ringspots followed by systemic necrotic spotting.

Serology
ToRSV, TRSV, PRMV and BBLMV are separate viruses that can be reliably distinguished and identified by serology. Immunodiffusion tests can be performed with sap expressed from symptomatic leaves of mechanically inoculated herbaceous hosts, but identification from field-grown vines is best made by immunoenzymatic procedures.

ELISA has been used successfully to detect ToRSV and PRMV from leaves of rooted cuttings and field-grown vines, with standard procedures (Ramsdell *et al.*, 1979; Gonsalves, 1980). The efficiency of detection is satisfactory, but extreme care must be taken in collecting the test samples because of the irregular distribution of the virus in the vines. Young leaves from different parts of the canopy and from suckers should be collected for assaying.

SANITATION
The same techniques used for GFLV and other European nepoviruses are applicable.

REFERENCES

Gilmer, R.M. & Uyemoto, J.K. 1972. Tomato ringspot virus in Baco noir grapevines in New York. *Plant Dis. Rep.*, 56: 133-135.

Gilmer, R.M, Uyemoto, J.K. & Kelts, J.L. 1970. A new grapevine disease induced by tobacco ringspot virus. *Phytopathology*, 60: 619-627.

Goheen, A.C. & Hewitt, W.B. 1962. Vein banding, a new virus disease of grapevines. *Am. J. Enol. Vitic.*, 13: 73-77.

Gonsalves, D. 1980. Detection of tomato ringspot virus in grapevines: irregular distribution of virus. *Proc. 7th Meet. ICVG*, Niagara Falls, NY, USA, 1980, p. 95-106.

Gooding, G.V. & Hewitt, W.B. 1962. Grape yellow vein: symptomatology, identification, and the association of a mechanically transmissible virus with the disease. *Am. J. Enol. Vitic.*, 13: 196-203.

Harrison, B.D. & Murant, A.F. 1977. *Nepovirus group*. Descriptions of Plant Viruses, No. 185.

Kew, UK, Commonw. Mycol. Inst./Assoc. Appl. Biol.

Martelli, G.P. & Taylor, C.E. 1989. Distribution of viruses and their nematode vectors. *Adv. Dis. Vector Res.,* 6: 151-189.

Piazzolla, P., Savino, V., Castellano, M.A. & Musci, D. 1985. A comparison of grapevine yellow vein virus and a grapevine isolate of tomato ringspot virus. *Phytopathol. Mediterr.,* 24: 44-50.

Ramsdell, D.C., Andrews, R.W., Gillet, J.M. & Morris, C.E. 1979. A comparison between enzyme-linked immunosorbent assay (ELISA) and *Chenopodium quinoa* for detection of peach rosette mosaic virus in Concord grapevines. *Plant. Dis. Rep.,* 63: 74-76.

Ramsdell, D.C. & Myers, R.L. 1974. Peach rosette mosaic virus, symptomatology and nematodes associated with grapevine degeneration in Michigan. *Phytopathology,* 64: 1174-1178.

Stace-Smith, R. & Ramsdell, D.C. 1987. Nepoviruses of the Americas. *Curr. Top. Vector Res.,* 5: 131-166.

Uyemoto, J.K., Taschenberg, E.F. & Hummer, D.K. 1977. Isolation and identification of a strain of grapevine Bulgarian latent virus in Concord grapevines in New York State. *Plant Dis. Rep.,* 61: 949-953.

Summary: grapevine decline (American nepoviruses) detection

GRAFT TRANSMISSION
Indicators
Grenache and Carignan (ToRSV yellow vein strain), Baco and Kober 5BB (ToRSV, type strain), Chardonnay (TRSV)
No. plants/test
3-5 rooted cuttings
Inoculum
Wood chips, single buds, bud sticks
Temperature
22-24°C
Symptoms
Chrome-yellow flecks along the veins (ToRSV, yellow vein); chlorotic blotches, rings, lines, foliar deformations (ToRSV, decline strain); leaf mottling, malformations, reduced growth (TRSV)

TRANSMISSION TO HERBACEOUS HOSTS
Diagnostic hosts
Phaseolous vulgaris (ToRSV and TRSV), *Cucumis sativus* (ToRSV and TRSV), *Chenopodium quinoa* (PRMV), *Nicotiana clevelandii* (BBLMV)
Inoculum
Tissues from young symptomatic leaves
Extraction
Grind in 2.5 percent aqueous nicotine
Temperature
Below 25°C
Symptoms
In *P. vulgaris* (ToRSV, TRSV), systemic necrosis of the topmost leaves in about 2 weeks;
In *C. sativus* (ToRSV, TRSV), chlorotic local lesions in 5-7 days followed by systemic mottling;
In *C. quinoa* (PRMV), faint chlorotic local lesions in about a week, systemic mottling and tip necrosis in 10-12 days;
In *N. clevelandii* (BBLM), local necrotic rings and systemic necrotic spotting in 10-12 days

OTHER TESTS
Serology (ELISA, ISEM)
Molecular hybridization (when probes are available)

FIGURE 39
Chlorotic rings and leaf deformation in a
Baco shoot infected by ToRSV
(Photo: D. Gonsalves)

FIGURE 40
Healthy Concord vine

FIGURE 41
ToRSV-infected Concord vine. Note the
markedly reduced crop

FIGURE 42
Typical yellow vein symptoms elicited by
the Californian strain of ToRSV

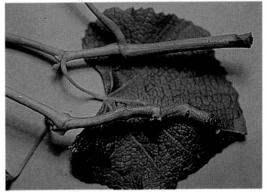

FIGURE 45
Irregular internodes and crooked canes consequent to
PRMV infection in a Concord vine
(Photo: D.C. Ramsdell)

FIGURE 43
Shoot of a field-grown
Chardonnay vine infected
by TRSV
(Photo: M.K. Corbett)

FIGURE 46
Straggling clusters in a PRMV-infected vine

FIGURE 44
Leaves from a declining Chardonnay vine infected by
TRSV. These symptoms are indistinguishable from those
induced by GFLV and other European nepoviruses
(Photo: M.K. Corbett)

FIGURE 47
Shock reaction of a Baco indicator to ToRSV
infection: chlorotic blotch
(Photo: D. Gonsalves)

FIGURE 48
Shock reaction of a Kober 5BB indicator to ToRSV
infection: chlorotic blotch with necrotizing margin
(Photo: D. Gonsalves)

FIGURE 49
Extremely severe symptoms in a Chardonnay
indicator inoculated with a TRSV source
(Photo: M.K. Corbett)

FIGURE 50
Phaseolus vulgaris with tip necrosis two weeks after inoculation with TRSV
(Photo: M.K. Corbett)

FIGURE 53
Plants of *Chenopodium quinoa* 9 days after inoculation with TRSV (left) and GFLV (right)
(Photo: M.K. Corbett)

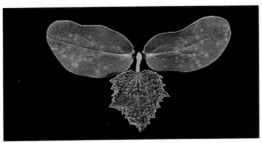

FIGURE 51
Cucumber plant a week after inoculation with TRSV showing chlorotic lesions on the cotyledons and systemic mottle
(Photo: M.K. Corbett)

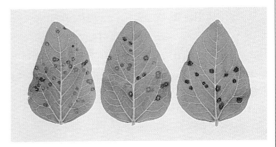

FIGURE 54
Necrotic lesions on leaves of cowpea inoculated with TRSV
(Photo: M.K. Corbett)

FIGURE 52
Chenopodium amaranticolor leaf with chlorotic lesions 9 days after inoculation with TRSV (left)
(Photo: M.K. Corbett)

FIGURE 55
Necrotic local lesions induced by PRMV in *C. quinoa*
(Photo: D.C. Ramsdell)

TRUE VIRUS DISEASES

Leafroll

G.P. Martelli

CAUSAL AGENTS

Grapevine leafroll may be induced by a complex of viruses, the majority of which belong to the closterovirus group. So far, at least five different long closteroviruses (i.e. with particle length ranging between 1 400 and 2 200 nm), denoted as grapevine leafroll-associated closteroviruses (GLRaV) I, II, III, IV and V, have been detected with varying degrees of consistency in infected vines (Gugerli, Brugger and Bovey, 1984; Hu, Gonsalves and Teliz, 1989; Zimmermann *et al.*, 1990). All have been characterized serologically and differ from one another. Two additional long closteroviruses which seem to be serologically unrelated to all of the above have also been detected recently (Zimmermann *et al.*, 1990). A shorter closterovirus (particles 800 nm long) called grapevine virus A (GVA) has also been found associated, though inconsistently, with the disease (Conti and Milne, 1985; Agran *et al.*, 1990).

All the above viruses are phloem-restricted and non-mechanically transmissible, except for GVA, some isolates of which can be transmitted with great difficulty by sap inoculation (Conti *et al.*, 1980; Agran *et al.*, 1990). There is mounting evidence that most if not all of these viruses, alone or in various combinations, are able to induce leafroll symptoms and therefore qualify as possible elicitors of the disease.

Described by Goheen, Harmon and Weinberger, 1958.

The aetiological role of a potyvirus serologically close to potato virus Y (Tanne *et al.*, 1977) is still undefined.

GEOGRAPHICAL DISTRIBUTION

The disease has been recorded from all major grapevine-growing areas of the world. The level of infection is often very high, so that cultivars grown in certain viticultural districts are totally diseased.

ALTERNATE HOSTS

No alternate hosts are known. Leafroll-associated closteroviruses have not been identified in any wild or cultivated plant species other than *Vitis* species and are not serologically related to any of the other known members of their taxonomic group.

FIELD SYMPTOMS

Affected vines may be smaller than healthy ones. Major external symptoms are downward rolling of the leaves accompanied by reddish-purple or yellow discolorations of the blades, depending on whether vines are red or white berried (Figures 56 and 57). Discoloured areas appear in the interveinal spaces of the lower leaves in early summer (Figure 58), becoming progressively stronger and extended so as to cover, with time, the whole foliar surface (Figure 59). The main veins may or may not retain the green colour in the advanced stages of the disease (Figure 60), and there is a difference in the hue, intensity and distribution of the reddish pigmentation over the leaf surface. In cases where the discoloration is particularly heavy, necrotic areas may develop in the interveinal tissues. Ripening of the fruits is affected. At harvest time, bunches are smaller than normal and may remain greenish or whitish (Figure 61). Clusters of certain red-fruited cultivars (e.g. Cardinal, Emperor) may become unmarketable because of the pale colouring of the berries.

As yet, it is not clear how extensively the variation in kind and intensity of symptoms shown in the field depends on varietal reactions or reflects differences in the type of agent(s) that cause specific responses. There are indications, however, that the presence of different closteroviruses is linked with differences in symptom expression. For example, GLRaV I seems to be related to a marked rolling of the leaf margins and a light reddish discoloration of the blades, whereas GLRaV III is more consistently associated with rolling of medium intensity and intense reddening with a deep purplish hue (Zimmermann, 1990).

NATURAL SPREAD

Although reports of leafroll spread within a vineyard are few, there is experimental evidence that spread is associated with pseudococcid mealybugs (Engelbrecht and Kasdorf, 1985). GVA can be acquired and transmitted by *Pseudococcus longispinus*, *Planococcus ficus* and *Planococcus citri* (Rosciglione *et al.*, 1983; Rosciglione and Castellano, 1985), whereas GLRaV III is transmitted from grape to grape by *P. longispinus* and *P. ficus* (Rosciglione and Gugerli, 1989; Tanne, Ben Dov and Raccah, 1989).

Long-distance spread is through infected propagating material of European grapes (budwood) and especially of American rootstocks (rooted cuttings), most of which are symptomless carriers of the disease. In certain cases, leafroll-

affected vines of *Vitis riparia* Gloire de Montpellier show a clear-cut rolling and yellowing of the leaves, which becomes evident in autumn.

DETECTION

Field symptoms are clearly expressed by red- and black-fruited European scions, although mild forms of the disease are known, which are less readily detected. The best time for field detection is in autumn when the foliar reddish-purple discolorations are at their peak (Figure 62). Detection in white-berried cultivars depends very much on the varietal reaction and possibly on the viral complex responsible for the infection. In these varieties the symptoms are rarely distinctive enough to be identified with certainty. Likewise, no field detection is possible in American rootstocks because of their lack of visible response to the disease, except for the above-mentioned instance of *V. riparia*. In these cases, laboratory detection is mandatory.

IDENTIFICATION
Indexing by graft transmission

Indicators for indexing are many, all belonging to black-fruited European varieties such as Mission, Pinot noir, Cabernet franc, Cabernet sauvignon and Barbera. The choice of the indicator may vary according to the conditions under which indexing is carried out. Thus, before establishing an indexing programme, it is advisable to run comparative graft transmission tests to assess the relative efficiency and reliability of symptom expression by each indicator.

The symptoms shown by the indicators are like those seen in the field: reddish discolorations on the interveinal areas of the leaves and downward rolling of their margins (Figure 63). In greenhouse indexing (green grafting) at 22°C, symptoms begin to appear about six weeks after grafting and are fully expressed in about three months (Walter *et al.*, 1990). Responses in field indexing (chip-budding, cleft or whip grafting) are slower, as they may take up to two years to show completely. However, severe forms of the disease induce clear-cut reactions in the first year of vegetation.

Baco 22A is recommended as a routine indicator for leafroll, since certain forms of the disease elicit in it severe stunting, yellowing and rolling of the leaves. These differential responses of woody indicators have not yet been correlated with the presence of specific single viruses or associations of viruses.

Transmission to herbaceous hosts

GVA was the first grapevine closterovirus transmitted by mechanical inoculation (Conti *et al.*, 1980). The transmission rate from expressed sap is extremely low and erratic, but with certain virus isolates transmission occurs more readily and can be enhanced by using partially purified preparations from grape leaves as inoculum.

Positive transmission is more consistently obtained, although not equally well with all GVA isolates, by means of pseudococcid vectors. Groups of five to ten crawling instars that have been allowed to feed on infected grapevine leaves are transferred, with excised pieces of the leaves they were feeding on, on to the leaves of young *Nicotiana clevelandii* and/or *Nicotiana benthamiana* seedlings. They are allowed to colonize the seedlings for a couple of weeks. The mealybugs may then be killed by insecticide treatment.

The herbaceous host range of GVA is extremely narrow, limited to *N. clevelandii* and *N. ben-thamiana*, both of which react with clearing and light yellowing of the secondary veins 10 to 12 days after infection (Figure 64).

Recent reports indicate that closteroviruses other than GVA can be mechanically transmitted

from diseased grapevines. These viruses have not yet been characterized, although it appears that their herbaceous host range is limited to *Nicotiana* species.

Micropurification

Micropurification is the procedure (see details in Part III) for obtaining virus preparations from grapevine tissues for multipurpose use (antiserum preparation, serology, immune electron microscopy, electrophoresis, etc.). The best sources of virus are petioles and main veins of old symptomatic leaves of European scions (8 to 10 g of tissues). Micropurification from American rootstock leaves, especially those containing *Vitis rupestris* plasma, gives inconsistent if not negative results (Boscia *et al.*, 1990). With American rootstocks, cortical tissue is a far better source of virus for micropurification.

Serology

Immunodiffusion in agar gel may not be used with closteroviruses, but media containing SDS (see Part III) for disrupting virus particles are suitable. In any case, the antigen must be a concentrated, partially purified virus preparation.

ELISA. Immunoenzymatic procedures, as described in Part III, can be used for detecting all known closteroviruses in field- or greenhouse-grown vines (Gugerli, Brugger and Bovey, 1984; Teliz *et al.*, 1987; Hu, Gonsalves and Teliz, 1989). Sources of antigen are fresh leaves, flowers, fruit, tendrils, bark and root tissues or wood shavings from dormant cuttings. The reactants can be polyclonal or monoclonal antibodies. Kits of both types of antibodies are commercially available for some closteroviruses (e.g. GVA and GLRaV I, as of 1990). Although the amount of plant material needed is as low as 40 mg (Teliz *et al.*, 1987), larger samples (up to 1 to 2 g) are commonly used. Tissues are ground in ordinary ELISA extraction buffer, with or without the addition of nicotine (1 to 2.5 percent), and are used thus or at a dilution of up to 1:50. Early field detection of viral antigens before symptoms appear on the foliage is possible from flower clusters and roots. ELISA reactions from basal leaves become consistently positive around blooming time. Viral antigens are evenly distributed in mature canes, which represent good antigen sources throughout the year (Teliz *et al.*, 1987). Negative and positive controls, chosen as indicated for GFLV, should always be included in the tests. For ELISA detection of leafroll-associated closteroviruses in American rootstocks, wood shavings are far superior to leaf tissues, which for *V. rupestris* and its hybrids give consistent negative results regardless of the time of sampling (Boscia *et al.*, 1990).

Immune electron microscopy (ISEM). GVA and other GLRaVs can be detected by ISEM following the procedure outlined in Part III. Although the starting material can be the same used for ELISA, cortical tissues taken after scraping off the bark are probably the best source of virus. Tissues are ground in three volumes of cold 2 percent PVP (MW 25 000 to 30 000) in 0.1 M phosphate buffer, pH 7, and the extract is used for observation (Conti and Milne, 1985).

Polyacrylamide gel electrophoresis (PAGE)

PAGE is used either for separation and characterization of double-stranded viral RNAs (dsRNA) or, in conjunction with Western blotting, for separation and identification of viral coat proteins. PAGE and Western blotting are applied as described in Part III to extracts from stem phloem tissues (Hu, Gonsalves and Teliz, 1989; Hu *et al.*, 1990). Both techniques disclose the presence of viral infections, and Western blotting also leads to the identification of the viral agent.

Molecular hybridization

Cloned cDNA probes have been prepared to genomic RNA of GVA, GLRaV III and the potyvirus serologically related to PVY. These probes have been successfully used for the identification of both viruses in field-grown grapevine leaf extracts denatured with formaldehyde (concentrated, partially purified preparations) or with 50 mM NaOH and 2.5 mM EDTA (grapevine sap) (Gallitelli, Savino and Martelli, 1985; Tanne, Naveh and Sela, 1989; Minafra, Russo and Martelli, 1990) and processed as described in Part III.

SANITATION

With many European red-berried cultivars visual selection helps reduce disease incidence. However, leafroll-free plants are best obtained by the following methods: prolonged heat treatment (60 to 120 days or more at 38°C) of grafted buds, or of whole plants with removal and rooting of shoot tips under mist (Goheen, 1977); heat treatment *in vitro* according to Galzy's method (Valat and Mur, 1976); micrografting (Bass and Legin, 1981) and *in vitro* meristem tip culture (Barlass *et al.*, 1982).

REFERENCES

Agran, M.K., Di Terlizzi, B., Boscia, D., Minafra, A., Savino, V., Martelli, G.P. & Askri, F. 1990. Occurrence of grapevine virus A (GVA) and other closteroviruses in Tunisian grapevines affected by leafroll disease. *Vitis*, 29: 49-55.

Barlass, M., Skene, K.G.M., Woodham, R.C. & Krake, L.R. 1982. Regeneration of virus-free grapevines using *in vitro* apical culture. *Ann. Appl. Biol.*, 101: 291-295.

Bass, P. & Legin, R. 1981. Thermothérapie et multiplication *in vitro* d'apex de vigne.

Application à la séparation ou à l'élimination de diverses maladies de type viral et à l'évaluation des dégâts. *C. R. Séances Acad. Agric. Fr.*, 67: 922-933.

Boscia D., Savino, V., Elicio, V., Jebahi, S.D. & Martelli, G.P. 1990. Detection of closteroviruses in grapevine tissues. *Proc. 10th Meet. ICVG*, Volos, Greece, 1990, p. 52-57.

Conti, M. & Milne, R.G. 1985. Closterovirus associated with leafroll and stem pitting in grapevine. *Phytopathol. Mediterr.*, 24: 110-113.

Conti, M., Milne, R.G., Luisoni, E. & Boccardo, G. 1980. A closterovirus from a stem-pitting diseased grapevine. *Phytopathology*, 70: 394-399.

Engelbrecht, D.J. & Kasdorf, G.G.F. 1985. Association of a closterovirus with grapevines indexing positive for grapevine leafroll and evidence for its natural spread in grapevine. *Phytopathol. Mediterr.*, 24: 101-105.

Gallitelli, D., Savino, V. & Martelli, G.P. 1985. The use of a spot hybridization method for the detection of grapevine virus A in the sap of grapevine. *Phytopathol. Mediterr.*, 24: 221-224.

Goheen, A.C. 1977. Virus and virus-like diseases of grapes. *Hortscience*, 12: 465-469.

Goheen, A.C., Harmon, F.N. & Weinberger, J.H. 1958. Leafroll (white emperor disease) of grapes in California. *Phytopathology*, 48: 51-54.

Gugerli, P., Brugger, J.J. & Bovey, R. 1984. L'enroulement de la vigne: mise en évidence de particules virales et développement d'une méthode immunoenzymatique pour le diagnostic rapide. *Rev. Suisse Vitic. Arboric. Hortic.*, 16: 299-304.

Hu, J.S., Gonsalves, D., Boscia, D. & Namba, S. 1990. Use of monoclonal antibodies to characterize grapevine leafroll associated closteroviruses. *Phytopathology*, 80 (in press).

Hu, J.S., Gonsalves, D. & Teliz, D. 1989. Characterization of closterovirus-like particles associated with grapevine leafroll disease. *J. Phytopathol.*, 128: 1-14.

Minafra, A., Russo, M. & Martelli, G.P. 1990. A cloned probe for the detection of grapevine virus A. *Proc. 10th Meet. ICVG*, Volos, Greece, 1990, p. 417-424.

Rosciglione, B. & Castellano, M.A. 1985. Further evidence that mealybugs can transmit grapevine virus A (GVA) to herbaceous hosts. *Phytopathol. Mediterr.*, 24: 186-188.

Rosciglione, B., Castellano, M.A., Martelli, G.P., Savino, V. & Cannizzaro, G. 1983. Mealybug transmission of grapevine virus A. *Vitis*, 22: 331-347.

Rosciglione, B. & Gugerli, P. 1989. Transmission of grapevine leafroll disease and an associated closterovirus to healthy grapevine by the mealybug *Planococcus ficus*. *Phytoparasitica*, 17: 63.

Tanne, E., Ben Dov, Y. & Raccah, B. 1989. Transmission of closterovirus-like particles by mealybugs (Pseudococcidae) in Israel. *Proc. 9th Meet. ICVG*, Kiryat Anavim, Israel, 1987, p. 71-73.

Tanne, E., Naveh, L. & Sela, I. 1989. Serological and molecular evidence for the complexity of the leafroll disease of grapevine. *Plant Pathol.*, 38: 183-189.

Tanne, E., Sela, I., Klein, M. & Harpaz, I. 1977. Purification and characterization of a virus associated with grapevine leafroll disease. *Phytopathology*, 67: 442-447.

Teliz, D., Tanne, E., Gonsalves, D. & Zee, F. 1987. Field serology of viral antigens associated with grapevine leafroll disease. *Plant Dis.*, 71: 704-709.

Valat, C. & Mur, T.G. 1976. Thermothérapie du Cardinal Rouge. *Prog. Agric. Vitic.*, 93: 200-204.

Walter, B., Bass, P., Legin, R., Martin, C., Vernoy, R., Collas, A. & Veselle, G. 1990. The use of a green-grafting technique for the detection of virus-like diseases of the grapevine. *J. Phytopathol.*, 128: 137-145.

Zimmermann, D. 1990. *La maladie de l'enroulement de la vigne: caractérisation de quatre particules virales de type closterovirus à l'aide d'anticorps polyclonaux et monoclonaux.* Ph.D. thesis. Univ. Louis Pasteur, Strasbourg. 256 pp.

Zimmermann, D., Bass, P., Legin, R. & Walter, B. 1990. Characterization and serological detection of four closterovirus-like particles associated with leafroll disease of grapevine. *J. Phytopathol.*, 130: 277-288.

Summary: leafroll detection

GRAFT TRANSMISSION

Indicators

Several cultivars of red-fruited *Vitis vinifera* (Pinot noir, Cabernet franc, Merlot, Barbera, Mission)

No. plants/test

3-5 rooted cuttings

Inoculum

Wood chips, single buds, bud sticks, shoot tips

Temperature

22°C (green grafting)

Symptoms

Rolling and reddening of the leaves in 4-6 weeks (green grafting) or 6-8 months to 2 years (field indexing)

TRANSMISSION TO HERBACEOUS HOSTS (GRAPEVINE VIRUS A)

Diagnostic hosts

Nicotiana clevelandii or *Nicotiana benthamiana*

Inoculum

Virus preparations micropurified from grapevine leaves or cortical tissues; viruliferous mealybugs; tissues from young leaves (with some virus isolates only)

Extraction

Grind leaf tissues in 2.5 percent aqueous nicotine

Temperature

Below 25°C

Symptoms

Systemic vein clearing and yellowing in 10-12 days

OTHER TESTS

Serology (ELISA, ISEM) and Western blot for closteroviruses for which antisera are available

Electrophoresis (PAGE) for dsRNA pattern

Molecular hybridization (GVA, GLRaV III)

FIGURE 56
Severe leafroll symptoms shown in autumn by a
red-fruited European grape cultivar

FIGURE 58
Incipient rolling and reddening of the leaves in a
vine infected by leafroll in spring

FIGURE 57
Yellowing and rolling of the leaves induced by leafroll
in a white-fruited European grape cultivar

FIGURE 59
Progressive reddish discolorations in leaves from
a vine affected by leafroll

FIGURE 60
Discoloured leaves with main veins retaining the
green colour

FIGURE 61
Pale-berried bunches from a leafroll-infected vine

FIGURE 63
Leafroll symptoms shown in autumn by an indicator
vine two years after graft inoculation

FIGURE 62
Red-fruited vines affected by leafroll are readily
identified in the field in autumn

FIGURE 64
Vein yellowing in *Nicotiana benthamiana* infected
with GVA

TRUE VIRUS DISEASES
Rugose wood complex

G.P. Martelli

CAUSAL AGENTS

Rugose wood is now considered to be a virus disease although its causal agent(s) has not yet been identified. The closterovirus GVA was originally isolated in Italy from a vine with a pitted trunk (Conti *et al.*, 1980). Later, inconsistent association of several serologically unrelated closteroviruses with wood pitting symptoms was reported from different countries. A tendentially more consistent association seems to exist between rugose wood *sensu lato* and GVA (Rosciglione and Gugerli, 1986; Zimmermann, 1990). However, some of these viruses, including GVA, are the same as those found in vines with leafroll disease. Thus, the assumption that the rugose wood complex is induced by one or more viruses is still largely based on its graft transmissibility and in part on its vector transmissibility. It should be noted, however, that a long closterovirus serologically differing from all those previously reported has recently been found in the original Californian source of corky bark (Namba *et al.*, 1991). Moreover, a short closterovirus (particles approximately 800 nm long) named grapevine virus B (GVB) was isolated by mechanical inoculation from corky bark-affected vines in Europe and North America (Boscia *et al.*, 1992). These recent findings strongly support the likelihood of a closterovirus aetiology for some of the components of the rugose wood complex.

Described by Savino, Boscia and Martelli, 1989.

GEOGRAPHICAL DISTRIBUTION

Rugose wood is widely distributed in most viticultural areas of the world.

ALTERNATE HOSTS

No host is known besides *Vitis* species.

FIELD SYMPTOMS

As defined in this handbook, rugose wood is a complex disease characterized by modifications of the woody cylinder. As specified below, four possibly different disorders can be recognized by indexing: rupestris stem pitting; corky bark; Kober stem grooving; and LN 33 stem grooving. Individual diseases cannot readily be distinguished in the field because of the absence of differential specific symptoms. In general, however, affected vines may be dwarfed and less vigorous than normal and may have delayed bud opening in spring. Some vines decline and die within a few years after planting. Grafted vines often show swelling above the bud union and a marked difference between the diameters of scion and rootstock (Figure 65). With certain cultivars, the bark of the scion above the graft union is exceedingly thick and corky and has a spongy texture and a rough appearance (Figure 66). The woody cylinder is typically marked by pits and/or grooves (Figure 67), which correspond to peg- and ridge-like protrusions on the cambial face of the bark (Figure 68). These alterations may occur on the scion (Figure 69), rootstock (Figure 70) or both (Figure 71), according to the cultivar/stock combination and possibly individual susceptibility. In most cases no specific symptoms are seen on the foliage, but bunches may be smaller and fewer than normal. Certain cultivars show foliage alterations similar to those induced by leafroll, i.e. rolling, yellowing or reddening of the leaf blades. These symptoms, when they occur, are more severe than those induced by ordinary forms of leafroll.

The ridges of the cortex consist of hyper-trophied rays extending from the bark into the functional xylem. Xylem strands are arranged irregularly, often being split into two or three smaller ones that run in diverging directions, thus conferring a fimbriate appearance to the woody cylinder. Parenchymatosis occurs in both xylem and phloem. These anatomical abnormalities originate from the altered behaviour of the vascular cambium.

NATURAL SPREAD

There is little doubt that rugose wood spreads primarily through infected propagative material. No field spread has been observed in Europe. However, there is evidence that in Mexico and Israel corky bark spreads in vineyards (Figure 72). The pseudococcid mealybug *Planococcus ficus* was reported to transmit corky bark experimentally from naturally diseased sources to the indicator LN 33 (Tanne, Ben Dov and Raccah, 1989).

DETECTION

Most rootstock/scion combinations express xylem symptoms in the field. To see the symptoms the bark must be removed; therefore, the best time for observation is during active vegetation (May to August in the Mediterranean area). The graft union is the first place to look for wood alterations. Sometimes the presence of pitting is obvious on the outer surface of the bark (Figure 73). If not, two parallel cuts 1 to 2 cm apart, and

a third cut across the first two, are made with a budding knife, and a strip of cortex is lifted to uncover the xylem (Figure 74). If no pits are seen, since their distribution may be irregular, new windows can be opened in the bark at different heights and locations above and below the graft union and around the whole circumference of the trunk. Infected vines may show no symptoms. In this case, field detection is not possible and diagnosis must rely on graft transmission tests.

IDENTIFICATION
Indexing by graft transmission

Rugose wood is transmitted by grafting, although sometimes without symptoms. Determination of the diseases forming the complex can be done using three differential indicators, i.e. *Vitis rupestris*, LN 33 and Kober 5BB.

Rupestris stem pitting. In *V. rupestris* this disease induces a distinctive basipetal pitting limited to a strip extending downward from the point of inoculation (Figure 75). LN 33 and Kober 5BB remain symptomless. Because of the peculiar localization of symptoms in *V. rupestris*, chip-bud grafting is recommended when indexing for rupestris stem pitting. In this case, the symptoms develop below the graft union toward the roots. Responses obtained by top grafting (Figure 76) are more difficult to detect. Wood symptoms appear two to three years after grafting.

Corky bark. This disease elicits grooving and pitting in all parts of the stem of *V. rupestris* and LN 33, but not in Kober 5BB. Furthermore, it induces proliferation of secondary phloem tissues of LN 33, giving rise to highly typical internodal swellings with a cracked surface (Figure 77). Infected LN 33 indicators are severely stunted and show early rolling and reddening of the leaves (Figure 78). Sometimes irregular yellow spots appear on the leaves of the spring flush (Figure 79). The canes ripen irregularly or not at all (Figure 80), and the vines may die within a year. Wood and cane symptoms may develop a few months after grafting, but usually they are fully expressed within two years. In green-grafted LN 33 indicators, symptoms on the canes may appear 20 to 40 days after inoculation.

Kober stem grooving. This disease induces a marked grooving on the stem of Kober 5BB (Figure 81) but no symptoms in *V. rupestris* and LN 33.

LN 33 stem grooving. Grooves of various lengths develop on the stem of LN 33 (Figure 82), much the same as with corky bark, but these are not accompanied by phloem proliferation leading to internodal swellings or by foliar discolorations. *Vitis rupestris* and Kober 5BB are symptomless.

Micropurification

Closteroviruses associated with rugose wood can be recovered by micropurification procedures as outlined in Part III. Petioles and main veins of old symptomatic leaves are good sources of virus, together with bark from dormant or green cuttings.

Serology

ELISA and ISEM can be applied using the procedure outlined for leafroll disease.

Polyacrylamide gel electrophoresis (PAGE)

Application of PAGE to RNA extracts from *in vitro* shoot-tip cultures of vines affected by rupestris stem pitting has led to the fairly consistent recovery of a small dsRNA (Monette, James and Godkin, 1989; Azzam, Gonsalves and Golino, 1991). The diagnostic value of these RNAs, however, requires further evaluation.

SANITATION

Visual sanitary selection carried out in established and aged (eight to ten year old) vineyards to a great extent helps eliminate rugose wood from selected clones. However, because of symptomless field infections, vines free of the complex are best obtained by heat treatment (more than 150 days at 38°C) and removal of shoot tips to be rooted or cultured *in vitro*.

REFERENCES

Azzam, O.I., Gonsalves, D. & Golino, D. 1991. Detection of dsRNA in grapevines showing symptoms of rupestris stem pitting disease and the variability encountered. *Plant Dis.*, 75: 690-694.

Beukman, E.F. & Goheen, A.C. 1966. Corky bark, a tumor-inducing virus of grapevines. *Proc. Int. Conf. Virus Vectors Perennial Hosts and* Vitis, 1965, p. 164-166. Div. Agric. Sci., Univ. Calif., Davis.

Boscia, D., Savino, V., Minafra, A., Namba, S., Elicio, V., Castellano, M.A., Gonsalves, D. & Martelli, G.P. 1992. Properties of a filamentous virus isolated from grapevines affected by corky bark. *Arch. Virol.*, (in press).

Conti, M., Milne, R.G., Luisoni, E. & Boccardo, G. 1980. A closterovirus from a stem pitting diseased grapevine. *Phytopathology*, 70: 394-399.

Goheen, A.C. 1988. Rupestris stem pitting. *In* R.G. Pearson and A.C. Goheen, eds. *Compendium of grape diseases*, p. 53. St Paul, MN, USA, Am. Phytopathol. Soc.

Graniti, A. & Martelli, G.P. 1966. Further observations on "legno riccio" (rugose wood), a graft transmissible stem pitting of grapevine. *Proc. Int. Conf. Virus Vector Perennial Hosts and* Vitis, 1965, p. 168-179. Div. Agric. Sci., Univ. Calif., Davis.

Monette, P.L., James, D. & Godkin, S.E. 1989. Double stranded RNA from rupestris stem pitting-affected grapevines. *Vitis*, 28: 137-144.

Namba, S., Boscia, D., Azzam, O., Maixner, M., Hu, J.S., Golino, D. & Gonsalves, D. 1991. Purification and properties of closterovirus-like particles associated with grapevine corky bark disease. *Phytopathology*, 81: 964-970.

Rosciglione, B. & Gugerli, P. 1986. Maladies de l'enroulement et du bois strié de la vigne: analyse microscopique et sérologique. *Rev. Suisse Vitic. Arboric. Hortic.*, 18: 207-211.

Savino, V., Boscia, D. & Martelli, G.P. 1989. Rugose wood complex of grapevine: can grafting to *Vitis* indicators discriminate between diseases? *Proc. 9th Meet. ICVG*, Kiryat Anavim, Israel, 1987, p. 91-94.

Tanne, E., Ben Dov, Y. & Raccah, B. 1989. Transmission of corky bark disease by the mealybug *Planococcus ficus*. *Phytoparasitica*, 17: 55.

Zimmermann, D. 1990. *La maladie de l'enroulement de la vigne: caractérisation de quatre particules virales de type closterovirus à l'aide d'anticorps monoclonaux et polyclonaux.* Ph.D. thesis. Univ. Louis Pasteur, Strasbourg. 256 pp.

Summary: rugose wood complex detection

GRAFT TRANSMISSION
Indicators
V. rupestris St George: rupestris stem pitting
LN 33: Corky bark
Kober 5BB: Kober stem grooving
LN 33: LN 33 stem grooving
No. plants/test
3-5 rooted cuttings
Inoculum
Wood chips or single buds (recommended for Rupestris stem pitting), bud sticks
Temperature
22°C (green grafting)
Symptoms
Basipetal pitting below grafted bud (stem pitting); internodal swellings and stem grooving in LN 33 (corky bark); stem grooving in Kober 55B only (Kober stem grooving); stem grooving in LN 33 only (LN 33 stem grooving)

OTHER TESTS
Serology (ELISA, ISEM) for the closterovirus associated with corky bark and GVA
Electrophoresis (PAGE) for dsRNA pattern
Molecular hybridization (GVA, GVB)

FIGURE 65
Swelling above the graft union in a vine affected by rugose wood. Note the difference in diameter between scion and rootstock and the rough appearance of the bark

FIGURE 67
Severe pitting of the trunk of a vine affected by rugose wood. The cortex has been removed to expose the altered woody cylinder

FIGURE 66
Corky appearance of the bark above the graft union in a vine affected by rugose wood

FIGURE 68
Peg- and ridge-like protrusions on the cambial face of the peeled cortex of a diseased vine correspond to pits and grooves on the woody cylinder

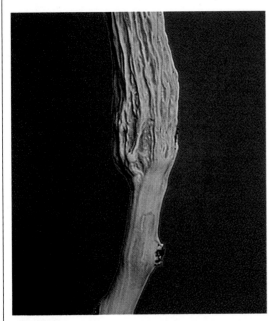

FIGURE 69
Symptoms of rugose wood showing only on the scion
of a diseased vine

FIGURE 71
Symptoms of rugose wood showing on both scion and
rootstock of a diseased vine

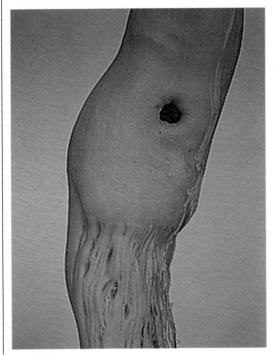

FIGURE 70
Symptoms of rugose wood showing only on the rootstock
of a diseased vine

FIGURE 72
Corky bark symptoms shown by an LN 33
vine naturally infected following mealybug
infestation in the field

FIGURE 73
Grooves can sometimes be seen
on the outer surface of the
cortex after removal of the bark

FIGURE 74
A window open in the cortex at the level of the graft union
to check the presence of rugose wood symptoms
(Photo: U. Prota)

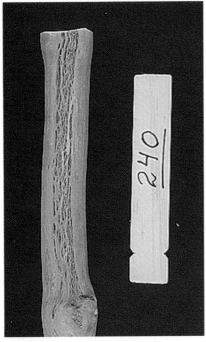

FIGURE 75
Typical strip of basipetal pitting extending
downward from the point of inoculation in a
V. rupestris indicator inoculated by
chip-budding with a source of rupestris
stem pitting disease
(Photo: A.C. Goheen)

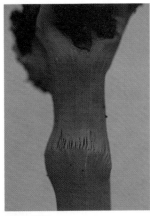

FIGURE 76
Rupestris stem pitting
symptoms in a *V. rupestris*
indicator inoculated by top
grafting. The pits are all around
and just below the graft union
(Photo: U. Prota)

FIGURE 77
Typical internodal swelling and cracking induced by
corky bark in LN 33

FIGURE 78
Stunting and leaf reddening of an LN 33 indicator
affected by a severe form of corky bark

FIGURE 79
Yellow spots sometimes appear in spring on leaves of
LN 33 graft-inoculated with corky bark sources

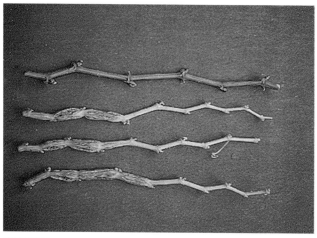

FIGURE 80
Canes from an LN 33 indicator showing classical corky bark
symptoms. A healthy mature cane (above) and immature canes
with internodal swellings from graft-inoculated vine

FIGURE 81
Stem grooving in a Kober 5BB indicator. Note absence
of symptoms on the scion
(Photo: B. Di Terlizzi)

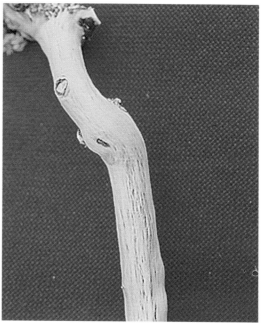

FIGURE 82
Stem grooving in an LN 33 indicator that did not show
secondary phloem proliferation. Note absence of
symptoms on the scion

TRUE VIRUS DISEASES
Yellow mottle
G.P. Martelli and J. Lehoczky

CAUSAL AGENT
The causal agent of yellow mottle is alfalfa mosaic virus (AMV), a mechanically transmissible virus with polymorphic particles (from quasi-isometric to bacilliform) measuring 28 to 58 x 18 nm and a tripartite genome.

GEOGRAPHICAL DISTRIBUTION
Yellow mottle has been reported only from central and eastern Europe (Germany, Switzerland, Czechoslovakia, Hungary and Bulgaria).

ALTERNATE HOSTS
AMV is a polyphagous virus infecting a great number of plant species in nature and artificially (Hull, 1969). The inoculum is widespread in temperate climates.

FIELD SYMPTOMS
Various patterns of yellow discolorations of the foliage characterize the disease. The spring growth shows in differing amounts yellowing of the leaf blades that does not extend to the veins, which remain green (Figure 83). Warm springs favour the development of symptoms. Faint yellow speckling, rings and lines (Figure 84) are typical summer responses of infected vines. Plant vigour and yield do not seem appreciably affected.

NATURAL SPREAD
AMV is transmitted in a non-persistent manner by aphids, which readily pass it from host to host. Although it is likely that grapevine infections originate from occasional inoculations by viruliferous aphids, there is no experimental evidence of this. The virus persists in propagative material and is perpetuated through it.

DETECTION
Field symptoms disclose the presence of diseased vines, especially in spring when symptom expression is at its peak.

IDENTIFICATION
Identification based on symptomatology is not possible because the symptons so closely resemble diseases induced by other viruses. It must be carried out with biological and laboratory procedures.

Indexing by graft transmission
AMV is readily transmitted by any of the graft inoculation procedures described in Part II. Chip-budding gives satisfactory results with several indicators: *Vitis rupestris*, Pinot noir, Siegfriedrebe, Mission, Chardonnay and Veltliner rouge précoce. The last two cultivars, because of their strong and consistent responses a few weeks after inoculation, are recommended as especially suitable indicators (Beczner and Lehoczky, 1981). The symptoms are much the same as those shown by naturally infected vines (Figure 85).

Described by Bercks, Lesemann and Querfurth, 1973.

Transmission to herbaceous hosts

AMV is readily transmitted by inoculation of sap to an extremely wide range of herbaceous hosts.

Leaf tissues ground in 0.07 M phosphate buffer, pH 7.2, containing 3 percent polyethylene glycol 6 000 (PEG), or in 2.5 percent aqueous nicotine are an excellent source of inoculum.

Diagnostic hosts are many:
- *Phaseolus vulgaris*. Necrotic local lesions (Figure 86) followed by systemic infection;
- *Chenopodium amaranticolor* and *Chenopodium quinoa*. Chlorotic/necrotic local lesions followed by systemic mosaic and top necrosis (Figure 87);
- *Vigna unguiculata*. Necrotic local lesions (Figure 88).

Serology

Immunodiffusion tests are easily performed with extracts from AMV-infected herbaceous hosts. ELISA was successfully used for the detection of AMV in naturally infected field-grown cv. Chardonnay vines in Hungary.

SANITATION

No information is available.

Summary: yellow mottle detection

GRAFT TRANSMISSION
Indicators
Vitis vinifera cv. Chardonnay and Veltliner rouge précoce
No. plants/test
3-5 rooted cuttings
Inoculum
Wood chips, single buds, bud sticks
Temperature
Field conditions
Symptoms
Yellow spots, rings and lines from 3-4 months after inoculation onward

TRANSMISSION TO HERBACEOUS HOSTS
Diagnostic hosts
Phaseolus vulgaris, Ocimum basilicum
Inoculum
Tissues from young symptomatic leaves
Extraction
Grind tissues in 0.07 M phosphate buffer, pH 7.2, containing 3 percent PEG or in 2.5 percent aqueous nicotine
Temperature
Below 25°C
Symptoms
In *P. vulgaris*, necrotic local lesions in 5-6 days and systemic mottling in 10-12 days;
In *O. basilicum*, systemic yellow blotches in about 2 weeks

OTHER TESTS
Serology (ELISA)

REFERENCES

Beczner, L. & Lehoczky, J. 1981. Grapevine disease in Hungary caused by alfalfa mosaic virus. *Acta Phytopathol. Acad. Sci. Hung.*, 16: 119-128.

Bercks, R., Lesemann, D. & Querfurth, G. 1973. Über der Nachweis des alfalfa mosaic virus in einer Weinrebe. *Phytopathol. Z.*, 76: 166-171.

Hull, R. 1969. Alfalfa mosaic virus. *Adv. Virus Res.*, 15: 365-433.

FIGURE 83
Bright yellow mottling induced by AMV infections
in spring

FIGURE 86
Necrotic local lesions induced
by AMV in *P. vulgaris*
(Photo: B. Walter)

FIGURE 84
Yellow spots and line patterns in naturally infected cv.
Chardonnay in summer

FIGURE 87
Systemic symptoms induced by AMV in
C. quinoa
(Photo: B. Walter)

FIGURE 85
Yellow line pattern in graft-
inoculated cv. Chardonnay
indicator

FIGURE 88
Necrotic local lesions induced
by AMV in *Vigna unguiculata*
(Photo: B. Walter)

TRUE VIRUS DISEASES
Line pattern
G.P. Martelli and J. Lehoczky

CAUSAL AGENT

Line pattern is caused by grapevine line pattern virus (GLPV), a possible member of the ilarvirus group. GLPV has polymorphic particles (quasi-spherical to bacilliform) 24 to over 100 nm in length and a multipartite genome.

GEOGRAPHICAL DISTRIBUTION

The disease has been reported only from Hungary, where it occurs in a few cultivars (Jubileum 75, Limberger and Oliver Irsai). Its incidence rarely exceeds 0.2 percent.

ALTERNATE HOSTS

No alternate host is known.

FIELD SYMPTOMS

Symptoms include bright yellow discolorations of the leaves forming marginal rings of variable size, scattered spots or blotches or maple-leaf line pattern (Figure 89). The line pattern is typically confined to the petiolar area or develops on the upper part of the leaf blade, roughly following its contour (Figure 90). The above symptoms are typical of the acute infection phase (shock symptoms). Chronically infected vines only show small yellow spots or flecking of the leaves (Figure 91), but vigour and yield are progressively reduced.

Described by Lehoczky *et al.*, 1990.

NATURAL SPREAD

No vector is known. The disease does not spread through the soil; it is perpetuated and transmitted through budwood and seeds (Lehoczky, Martelli and Lazar, 1992).

DETECTION

Field symptoms are fairly obvious in newly infected vines. Chronic symptoms are more difficult to detect and require careful observation of the leaves in summer.

IDENTIFICATION

Shock symptoms of line pattern resemble those of AMV-induced yellow mottle. Chronic symptoms resemble those elicited by any of the several chromogenic virus strains that infect grapevines. Thus proper identification relies on biological and laboratory tests.

Indexing by graft transmission

Line pattern is transmissible by grafting to a number of European grape cultivars. The field syndrome is reproducible by graft inoculation (chip-budding, cleft- or whip-grafting) to Jubileum 75, but symptoms are slow to appear.

Transmission to herbaceous hosts

GLPV is readily transmitted by inoculation of sap to a moderately wide range of herbaceous hosts. Leaf tissues ground in 0.07 M phosphate buffer, pH 7.2, containing 3 percent polyethylene glycol 6 000 (PEG) are a satisfactory source of inoculum.

Diagnostic hosts are:

- *Chenopodium amaranticolor* and *Chenopodium quinoa*. Chlorotic local lesions followed by systemic mosaic, distortion and apical necrosis (Figure 92);
- *Cucumis sativus*. Chlorotic local lesions and systemic chlorotic mottle;
- *Nicotiana glutinosa*. Chlorotic local lesions (Figure 93) followed by systemic mottling and necrosis;
- *Gomphrena globosa*. Chlorotic local lesions (Figure 94) turning reddish with age and mild systemic mottling.

Serology

Immunodiffusion tests are readily performed with extracts from GLPV-infected herbaceous hosts. ELISA and ISEM are likely to be applicable for virus detection in field samples, but they have not yet been tested.

SANITATION

No information is available.

REFERENCES

Lehoczky, J., Boscia, D., Burgyan, J., Castellano, M.A., Beczner, L. & Farkas, G. 1990. Line pattern, a novel virus disease of the grapevine in Hungary. *Proc. 9th Meet. ICVG*, Kiryat Anavim, Israel, 1987, p. 23-30.

Lehoczky, J., Martelli, G.P. & Lazar, J. 1992. Seed transmission of grapevine line pattern virus. *Phytopathol. Mediterr.*, 31: 115-116.

Summary: line pattern detection

GRAFT TRANSMISSION
Indicator
Vitis vinifera cv. Jubileum 75
No. plants/test
3-5 rooted cuttings
Inoculum
Wood chips, single buds, bud sticks
Temperature
Field conditions
Symptoms
Yellow rings and line patterns in the second year after grafting

TRANSMISSION TO HERBACEOUS HOSTS
Diagnostic hosts
Chenopodium amaranticolor, C. quinoa, Nicotiana glutinosa
Inoculum
Tissue from young symptomatic leaves
Extraction
Grind in 0.07 M phosphate buffer, pH 7.2, with 3 percent PEG or in 2.5 percent aqueous nicotine
Temperature
Below 25°C
Symptoms
In *Chenopodium* species, chlorotic local lesions in about a week, systemic mottle and apical necrosis in 10-12 days; In *N. glutinosa*, chlorotic local lesions in 5-7 days, systemic mottling and necrosis in 10-12 days

FIGURE 91
Scattered minute yellow spots
associated with chronic line
pattern infection

FIGURE 89
Distribution of line pattern symptoms in a
newly infected shoot of cv. Jubileum 75

FIGURE 90
Yellow rings and lines typically associated with
grapevine line pattern disease

FIGURE 92
Distortion, mosaic and apical necrosis
induced by GLPV in *C. quinoa*

FIGURE 93
Chlorotic local lesions in a *N. glutinosa* leaf
inoculated with GLPV

FIGURE 94
Young chlorotic lesions induced by GLPV in *G. globosa*

TRUE VIRUS DISEASES
Fleck
G.P. Martelli

CAUSAL AGENT
The causal agent of fleck is grapevine fleck virus (GFkV), a phloem-limited, non-mechanically transmissible isometric virus containing a single molecule of ssRNA (Boscia *et al.*, 1991).

GEOGRAPHICAL DISTRIBUTION
Fleck was first identified as a disease in its own right in California by Hewitt and co-workers (1972). It has been recorded from a great many countries and is now thought to have a worldwide distribution.

ALTERNATE HOSTS
No alternate host is known.

FIELD SYMPTOMS
Field symptoms are absent in all European scions and in most American rootstock species and hybrids, in which the disease is latent. Symptoms are clearly shown by naturally infected, self-indexing *Vitis rupestris* St George. These consist of clearing of the veins of third and fourth order producing localized translucent spots, which are best seen by holding the leaves against the light (Figure 95). Leaves with intense flecking are wrinkled and twisted and may curl upward. Severe strains also induce varying degrees of stunting.

NATURAL SPREAD
No vector is known. Seed transmission does not occur. Although the disease can experimentally be transmitted through dodder (*Cuscuta* spp.), (Woodham and Krake, 1983), spread is primarily through infected propagating material.

DETECTION
Detection in the field is only possible in infected, symptomatic *V. rupestris*.

IDENTIFICATION
Indexing by graft transmission
Grafting to *V. rupestris* St George produces the symptoms described above. In the field or greenhouse, symptoms appear five to six weeks after inoculation. They appear earlier (four weeks) when the indicators are grown at 22°C under continuous illumination in a growth chamber (Mink and Parsons, 1977). Mild strains may induce positive reactions in the year after grafting. The symptoms tend to fade away in summer, so readings should be made within eight to ten weeks after inoculation. Identification of fleck symptoms may be difficult if indexed vines also carry grapevine fanleaf. In this case, early readings as soon as foliar flecks appear may be useful for distinguishing between the two diseases.

Serology
ELISA and ISEM can be used to detect the causal virus in roots, leaves and cortex (bark scraping) of infected vines, even when these do not show

---

Described by Hewitt *et al.*, 1972.

symptoms, as is the case with all *V. vinifera* cultivars.

SANITATION

Heat treatment at 38°C for not less than 60 days, removal and rooting under mist of shoot tips 5 to 8 mm long eliminates the disease from about 90 percent of the explants. Complete elimination of the causal agent is achieved by culturing 1 mm long explants *in vitro* at 30°C with a 15 hour photoperiod (Barlass *et al.*, 1982).

REFERENCES

Barlass, M., Skene, K.G.M., Woodham, R.C. & Krake, L.R. 1982. Regeneration of virus-free grapevines using *in vitro* apical culture. *Ann. Appl. Biol.*, 102: 291-295.

Boscia, D., Martelli, G.P., Savino, V. & Castellano, M.A. 1991. Identification of the agent of grapevine fleck disease. *Vitis*, 30: 97-105.

Hewitt, W.B., Goheen, A.C., Cory, L. & Luhn, C. 1972. Grapevine fleck disease, latent in many varieties, is transmitted by graft inoculation. *Ann. Phytopathol.*, hors sér., p. 43-47.

Mink, G.I. & Parsons, J.L. 1977. Procedures for rapid detection of virus and virus-like diseases of grapevine. *Plant Dis. Rep.*, 61: 567-571.

Woodham, R.C. & Krake, L.R. 1983. Investigations on transmission of grapevine leafroll, yellow speckle and fleck diseases by dodder. *Phytopathol. Z.*, 106: 193-198.

Summary: fleck detection

GRAFT TRANSMISSION
Indicator
Vitis rupestris St George
No. plants/test
3-5 rooted cuttings
Inoculum
Wood chips, single buds, bud sticks, shoot tips
Temperature
22°C (green grafting or growth chamber)
Symptoms
Clearing of the veinlets in 4-6 weeks according to growing conditions

OTHER TESTS
Serology (ELISA, ISEM)

FIGURE 95
Clearing of the veinlets of *V. rupestris* typical of fleck disease

TRUE VIRUS DISEASES
Ajinashika disease
S. Namba and G.P. Martelli

CAUSAL AGENT
This disease was reported to be caused by the concurrent infection of leafroll and fleck (Terai, 1990). However, an isometric, phloem-limited, non-mechanically transmissible RNA virus about 25 nm in diameter, consistently found in affected vines, is now suspected as the causal agent. This virus differs and is serologically distinct from the agent of fleck (Namba *et al.*, 1991).

GEOGRAPHICAL DISTRIBUTION
This disease has been reported only from Japan.

ALTERNATE HOSTS
No alternate hosts are known. The disease is apparently restricted to *Vitis vinifera* cv. Koshu.

FIELD SYMPTOMS
Infected vines do not show appreciable symptoms on the foliage or apparent reduction of vigour and yield (Figure 96). The berries, however, are pale-coloured (Figure 97) and have a low sugar content, which makes the crop unmarketable. This condition gives the name to the disease, which is Japanese for "unpalatable fruit with low sugar content". American rootstock hybrids are infected without showing symptoms.

NATURAL SPREAD
No vector is known. Spread is primarily through infected propagating material.

———
Described by Namba *et al.*, 1986.

DETECTION
Pale colouring and low sugar content of the berries disclose the presence of diseased vines. Symptoms are best seen at harvesting time.

IDENTIFICATION
Indexing by graft transmission
The disease is readily transmitted by grafting to healthy cv. Koshu vines. The symptomatology is reproduced two to three years after grafting.

Serology
ELISA and ISEM can be used to detect the suspected causal virus in various organs of infected vines. Green shoots and matured fruit cores are the best antigen sources.

SANITATION
Prolonged heat treatment as used for leafroll and fleck eliminates the disease.

REFERENCES

Namba, S., Boscia, D., Yamashita, S., Tsuchizaki, T. & Gonsalves, D. 1991. Purification and properties of spherical virus particles associated with grapevine ajinashika disease. *Plant Dis.*, 75: 1249-1253.

Namba, S., Iwanami, T., Yamashita, S., Doi, Y. & Hatamoto, M. 1986. Three phloem-limited viruses of grapevine: direct fluorescence detection. *Food Fert. Technol. Cent. Taiwan Tech. Bull.*, 92: 1-17.

Terai, Y. 1990. Ajinashika disease: a combined effect of grapevine leafroll and grapevine fleck viruses on sugar content in the Japanese grape cultivar Koshu. *Proc. 10th Meet. ICVG*, Volos, Greece, 1990, p. 67-70.

Summary: ajinashika disease detection

GRAFT TRANSMISSION
Indicator
Vitis vinifera cv. Koshu
No. plants/test
3-5 rooted cuttings
Inoculum
Wood chips, single buds, bud sticks
Temperature
Field conditions
Symptoms
Same as in the field in 2-3 years

OTHER TESTS
Serology (ELISA, ISEM)

FIGURE 96
A cv. Koshu vineyard with vines affected by ajinashika disease,
showing pale-coloured bunches

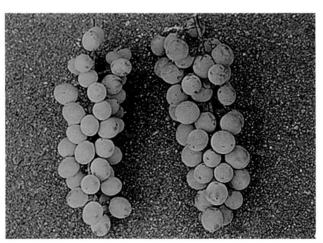

FIGURE 97
Bunches of cv. Koshu from a diseased (left) and a healthy (right) vine.
Note the whitish colour and the smaller berries of the diseased bunch

TRUE VIRUS DISEASES
Grapevine stunt
S. Namba and G.P. Martelli

CAUSAL AGENT

An isometric, phloem-limited, non-mechanically transmissible virus about 25 nm in diameter, consistently associated with diseased vines, is regarded as the possible agent. This virus is serologically unrelated to the putative agent of ajinashika disease (Namba *et al.*, 1986).

GEOGRAPHICAL DISTRIBUTION

The disease has been reported only from Japan.

ALTERNATE HOSTS

None are known. The disease is apparently restricted to *Vitis vinifera* cv. Campbell Early.

FIELD SYMPTOMS

Spring vegetation is delayed, internodes are short and leaves are small and curled and sometimes have scorched margins (Figure 98). Inflorescences are undersized, fruit-setting is impaired and bunches are few and shelled. Severely infected vines may be unfruitful. Summer recovery occurs so that the newly produced vegetation is apparently normal. Symptom expression is stronger in young (one to four year old) vines.

NATURAL SPREAD

The disease is transmitted in nature by the grapevine leafhopper *Arboridia apicalis* (Figure 99). Spread also occurs through infected propagating material.

DETECTION

Affected vines are readily identified in the field, especially in spring when symptom expression is at its peak.

IDENTIFICATION

Indexing by graft transmission

The disease is readily transmitted by grafting to cv. Campbell Early, in which the field syndrome is reproduced in about a year.

Transmission by vectors

Adults and larvae of *A. apicalis* transmit the disease agent to cv. Campbell Early vines with acquisition and inoculation periods of five and seven days, respectively. Symptoms appear about a year after feeding of viruliferous insects.

Serology

An antiserum for ELISA testing is being developed.

SANITATION

The virus was eliminated from up to 40 percent of grapevine explants after treatment for five months at two temperature regimes (30°C for 10 hours and 40°C for 14 hours).

Described by Namba *et al.*, 1986.

REFERENCE

Namba, S., Iwanami, T., Yamashita, S., Doi, Y. & Hatamoto, M. 1986. Three phloem-limited viruses of grapevine: direct fluorescence detection. *Food Fert. Technol. Cent. Taiwan Tech. Bull.*, 92: 1-17.

Summary: grapevine stunt detection

GRAFT TRANSMISSION
Indicator
Vitis vinifera cv. Campbell Early
No. plants/test
3-5 rooted cuttings
Inoculum
Wood chips, single buds, bud sticks
Temperature
Field conditions
Symptoms
Same as in the field in about a year

OTHER TESTS
Serology (ELISA is being developed)

FIGURE 98
Stunt symptoms in spring. Note poor growth and curling of
the leaves

FIGURE 99
Adult of *Arboridia apicalis*, the leafhopper
vector of grapevine stunt

TRUE VIRUS DISEASES
Roditis leaf discoloration
I.Ch. Rumbos, A.D. Avgelis and G.P. Martelli

CAUSAL AGENT
Diseased vines are doubly infected by GFLV and carnation mottle carmovirus (CarMV) (Avgelis and Rumbos, 1990). CarMV is a virus with isometric particles about 30 nm in diameter and a monopartite RNA genome.

GEOGRAPHICAL DISTRIBUTION
The disease has been reported only from Greece.

ALTERNATE HOSTS
CarMV has a fairly narrow range of natural hosts, but experimentally it infects a wide spectrum of botanical species (Hollings and Stone, 1970).

FIELD SYMPTOMS
Field symptoms include yellowish and/or reddish discolorations of tissues along the veins, interveinal areas or variously extended sectors of the leaf blade, especially near the petiole (Figures 100 and 101). Leaves are deformed, usually in correspondence to discoloured sectors. Bunches are reduced in number and size and have a low sugar content. The symptoms appear in spring and persist through the vegetative season.

NATURAL SPREAD
No vector is known. Spread is through infected propagating material.

DETECTION
Diseased vines are readily identified in the field because of their yellow foliage.

IDENTIFICATION
Indexing by graft transmission
The disease is readily transmitted by grafting to *Vitis vinifera* cv. Mission. Symptoms appear within the first year after grafting and are similar to those observed in the field.

Transmission to herbaceous hosts
CarMV and GFLV are both mechanically transmissible to a fairly wide range of herbaceous hosts. Diagnostic hosts for the grapevine isolate of CarMV are:
- *Chenopodium quinoa*. Chlorotic local lesions and systemic mottling;
- *Gomphrena globosa*. Reddish local lesions, systemic mottling and leaf deformation (Figure 102);
- *Nicotiana clevelandii*. Necrotic rings and spots in the inoculated leaves (Figure 103).

Serology
ELISA can be used for GFLV detection in naturally infected vines and may also be applicable for the detection of CarMV.

SANITATION
No information is available.

Described by Rumbos and Avgelis, 1989.

REFERENCES

Avgelis, A.D. & Rumbos, I.Ch. 1990. Carnation mottle virus isolated from vines affected by Roditis leaf discoloration. *Proc. 10th Meet. ICVG,* Volos, Greece, 1990, p. 437-443.

Hollings, M. & Stone, O.W. 1970. *Carnation mottle virus.* Descriptions of Plant Viruses, No. 7. Kew, UK, Commonw. Mycol. Inst./Assoc. Appl. Biol.

Rumbos, I.Ch. & Avgelis, A.D. 1989. Roditis leaf discoloration. A new disease of grapevine: symptomatology and transmission to indicator plants. *J. Phytopathol.,* 125: 274-278.

Summary: Roditis leaf discoloration detection

GRAFT TRANSMISSION
Indicator
Vitis vinifera cv. Mission
No. plants/test
3-5 rooted cuttings
Inoculum
Wood chips, single buds, bud sticks
Temperature
Field conditions
Symptoms
Same as in the field in about a year

TRANSMISSION TO HERBACEOUS HOSTS
Diagnostic hosts
Chenopodium quinoa, Gomphrena globosa
Inoculum
Tissue from young symptomatic leaves or root tips
Extraction
Grind in 2.5 percent aqueous nicotine
Temperature
Below 25°C
Symptoms (CarMV)
In *C. quinoa,* chlorotic local lesions, systemic mottling; In *G. globosa,* reddish local lesions, systemic mottling, leaf deformation

OTHER TESTS
Serology (ELISA) for GFLV and possibly also for CarMV

FIGURE 100
Sectorial discolorations and vein yellowing
in leaves of cv. Roditis vines affected by
leaf discoloration

FIGURE 102
Local necrotic lesions and deformation of the upper
leaves in *G. globosa* infected by the grapevine
isolate of CarMV

FIGURE 101
Reddish and yellow discolorations of the veins and
margins of cv. Roditis leaves affected by leaf
discoloration

FIGURE 103
Necrotic spots and rings on a *N. clevelandii* leaf
inoculated with the grapevine isolate of CarMV

VIROID DISEASES
Yellow speckle
G.P. Martelli

Although six viroids have been recovered so far from grapevines (see Table 3), (Sano *et al.*, 1985; Flores *et al.*, 1985; Semancik, Rivera-Bustamante and Goheen, 1987; Rezaian, Koltunow and Krake, 1988; Puchta, Ramm and Sanger, 1989; Semancik and Szychowski, 1990), four are not known to induce disease. These viroids, i.e. hop stunt viroid (HSVd), citrus exocortis viroid (CEVd), Australian grapevine viroid (AGVd) and grapevine viroid – cucumber (GVd-c), have been found in vines without symptoms or with symptoms of different virus and/or virus-like diseases. The only disease for which a viroidal aetiology has reasonably been established is yellow speckle disease.

CAUSAL AGENT
The causal agents of yellow speckle disease are grapevine yellow speckle viroids 1 and 2 (GYSVd-1 and GYSVd-2), viroids with chain lengths of 367 and 363 nucleotides, respectively (Koltunow and Rezaian, 1989).

GEOGRAPHICAL DISTRIBUTION
GYSVds may have a worldwide distribution. They have been found in vines originating from many different countries of all continents (Semancik, Rivera-Bustamante and Goheen, 1987; Szychowski, Goheen and Semancik, 1988; Minafra, Martelli and Savino, 1990).

Described by Taylor and Woodham, 1972.

ALTERNATE HOSTS
No natural hosts are known other than *Vitis* species.

FIELD SYMPTOMS
Yellow speckle is an elusive disease whose outward expression is conditioned by climatic and possibly varietal factors. Symptoms, when shown, consist of a few to many minute chrome yellow spots or flecks scattered over part or all of the leaf surface (Figures 104 and 105) or gathering along the main veins to give a vein banding pattern (Figure 106).

GYSVd-induced vein banding is very similar, if not identical, to the symptoms of a disease also known by the name of vein banding (Figure 107), which has long been regarded as part of the fanleaf degeneration complex (Goheen and Hewitt, 1962). Although vein banding may show in GFLV-free vines, it is more often associated with GFLV infections. In fact, it has been suggested that the presence of GFLV enhances the expression of GYSVd in the form of vein banding patterns (Krake and Woodham, 1983). Similar enhancement may occur in vines concurrently infected by grapevine chrome mosaic virus (Figure 108). Unlike GFLV-induced yellow discolorations, the symptoms of yellow fleck appear in the height of summer on a limited number of mature leaves, and they persist for the rest of the vegetating season.

Yellow speckle symptoms may also show concurrently with symptoms of other diseases such as, for instance, leafroll (Figure 109).

NATURAL SPREAD

No vector is known. Natural dissemination takes place by mechanical inoculation through surface-contaminated cutting tools during management operations (pruning and propagation); by graft transmission; and by distribution of infected propagating material. The absence of symptoms in most European scion varieties and all American rootstocks greatly facilitates inadvertent viroid dissemination, making viroid dispersal virtually impossible to prevent. None of the grapevine viroids is known to be seed-transmitted.

DETECTION

Detection is based on observation of field symptoms. A search for symptoms (especially vein banding) is best made in late summer. For disease detection in locations where symptoms do not develop under field conditions, a self-indexing procedure may be used. Freshly collected or cold-stored grapevine cuttings are placed in a growth chamber at 32°C under continuous illumination (fluorescent tubes and incandescent bulbs giving 10 000 lux at the level of vegetation) and are left to root and grow for four weeks (Mink and Parsons, 1975). Translucent chlorotic spots develop along the major veins and interveinal tissues of GYSVd-infected vines 10 to 20 days after the growth starts.

IDENTIFICATION
Indexing by graft transmission

Indicators for yellow speckle are European cultivars Mataro and Mission. Symptoms are like those seen in naturally infected vines in the field but are equally erratic. Under most indexing conditions indicator responses are unreliable. If chip-bud-inoculated LN 33 cuttings are immediately placed in a growth chamber at 32°C for three weeks, then transferred to a cooler environment (18 to 20°C) under continuous illumination at 10 000 lux, they may develop yellow-green to chrome-yellow foliar spots on a few leaves, which sometimes give a vein banding pattern.

Transmission to herbaceous hosts

GYSVd-1 was reported to be mechanically transmissible to cucumber cv. Suyo, which it infects without symptoms (Semancik, Rivera-Bustamante and Goheen, 1987). Other grapevine viroids are pathogenic to tomato cv. Rutgers (HSVd, AGVd and CEVd), cucumber (HSVd, AGVd, GVd-c) and *Gynura aurantiaca* (CEVd), in which they induce visible symptoms.

Polyacrylamide gel electrophoresis (PAGE)

Bioassays are unreliable, so PAGE is used for detection and tentative identification of GYSVds and other grapevine viroids. The procedures for extraction, concentration and electrophoresis are as reported in Part III. Identification is made on the basis of electrophoretic mobility compared with that of reference viroids of known size (e.g. HSVd and CEVd).

Molecular hybridization

Molecular probes to grapevine viroids have been made and have proved useful for specific identification (Rezaian, Koltunow and Krake, 1988; Semancik and Szychowski, 1990).

SANITATION

Heat treatment does not free grapevine explants from yellow speckle. Sanitation is obtained by regenerating plantlets from fragmented shoot apices grown *in vitro* at temperatures between 20°C (night) and 27°C (day) (Barlass *et al.*, 1982) or by culturing 0.1 to 0.2 mm long shoot tips at 25 to 27°C. The latter method also eliminates HSVd (Duran-Vila, Juarez and Arregui, 1988).

REFERENCES

Barlass, M., Skene, K.G.M., Woodham, R.C. & Krake, R.L. 1982. Regeneration of virus free grapevines using *in vitro* apical culture. *Ann. Appl. Biol.*, 101: 291-295.

Duran-Vila, N., Juarez, J. & Arregui, J.M. 1988. Production of viroid-free grapevines by shoot tip culture. *Am. J. Enol. Vitic.*, 39: 217-220.

Flores, R., Duran-Vila, N., Pallas, V. & Semancik, J.S. 1985. Detection of viroid and viroid-like RNA from grapevine. *J. Gen. Virol.*, 66: 2095-2102.

Goheen, A.C. & Hewitt, W.B. 1962. Vein banding, a new virus disease of grapevines. *Am. J. Enol. Vitic.*, 13: 73-77.

Koltunow, A.M. & Rezaian, M.A. 1989. Grapevine viroid 1B, a new member of the apple scar skin viroid group, contains the left terminal region of tomato planta macho viroid. *Virology*, 170: 575-578.

Krake, R.L. & Woodham, R.C. 1983. Grapevine yellow speckle agent implicated in the aetiology of vein banding disease. *Vitis*, 22: 40-50.

Minafra, A., Martelli, G.P. & Savino, V. 1990. Viroids of grapevines in Italy. *Vitis*, 29: 173-182.

Mink, G.I. & Parsons, J.L. 1975. Rapid indexing procedures for detecting yellow speckle disease in grapevines. *Plant Dis. Rep.*, 59: 869-872.

Puchta, H., Ramm, K. & Sanger, H.L. 1988. Nucleotide sequence of a hop stunt viroid isolate from the German grapevine cultivar "Riesling". *Nucl. Acids Res.*, 16: 2730.

Rezaian, M.A., Koltunow, A.M. & Krake, L.R. 1988. Isolation of three viroids and a circular RNA from grapevine. *J. Gen. Virol.*, 69: 413-422.

Sano, T., Uyeda, I., Shikata, E., Meshi, T., Ohno, T. & Okado, Y. 1985. A viroid-like RNA isolated from grapevine has high sequence homology with hop stunt viroid. *J. Gen. Virol.*, 66: 333-338.

Semancik, J.S., Rivera-Bustamante, R. & Goheen, A.C. 1987. Widespread occurrence of viroid-like RNA in grapevine. *Am. J. Enol. Vitic.*, 38: 35-40.

Semancik, J.S. & Szychowski, J.A. 1990. Comparative properties of viroids of grapevine origin isolated from grapevines and alternate hosts. *Proc. 10th Meet. ICVG*, Volos, Greece, 1990, p. 270-278.

Szychowski, J., Goheen, A.C. & Semancik, J.S. 1988. Mechanical transmission and rootstock reservoirs as factors in the widespread distribution of viroids in grapevines. *Am. J. Enol. Vitic.*, 39: 213-216.

Taylor, R.H. & Woodham, R.C. 1972. Grapevine yellow speckle: a newly recognized graft-transmissible disease of *Vitis*. *Aust. J. Agric. Res.*, 23: 447-452.

Summary: yellow speckle detection

GRAFT TRANSMISSION
Indicators
Vitis vinifera cv. Mission or LN 33
No. plants/test
3-5 rooted cuttings
Inoculum
Wood chips, single buds
Temperature
32°C for 3 weeks, then 18-20°C (growth chamber). Above 25°C in the open field
Symptoms
In LN 33, yellow-green to chrome-yellow spots and/or vein banding;
In Mission, yellow flecks and/or vein banding

OTHER TESTS
Sequential electrophoresis (sPAGE)
Molecular hybridization

FIGURE 104
Scattered yellow spots in a
European grape leaf, typical
of yellow speckle infection

FIGURE 107
Vein banding symptoms in a
vine doubly infected with yellow
speckle viroid and grapevine
fanleaf virus

FIGURE 105
Symptoms as in Figure 104, but with
more intense yellow speckling

FIGURE 108
Vein banding symptoms in a vine
doubly infected with yellow
speckle viroid and grapevine
chrome mosaic virus
(Photo: J. Lehoczky)

FIGURE 106
Strong yellow speckle symptoms with speckles
tending to gather along the main veins

FIGURE 109
Symptoms of yellow speckle and leafroll appearing
in autumn on the same vine

VIRUS-LIKE DISEASES
Enation disease

G.P. Martelli

CAUSAL AGENT
The causal agent is unknown.

GEOGRAPHICAL DISTRIBUTION
Although known to occur in Italy since the beginning of the century (Graniti, Martelli and Lamberti, 1966), enation disease was recognized as a specific disorder by Hewitt (1954) in California. It occurs in most European countries, Israel, the United States, Venezuela, South Africa, New Zealand and Australia.

ALTERNATE HOSTS
No alternate host is known.

FIELD SYMPTOMS
Field symptoms include delayed bud breaking, slow bushy growth of the shoots in the initial stages of vegetation and presence of laminar or cup-shaped outgrowths (enations) on the underside of the eight to ten leaves at the base of the shoot (Figures 110 and 111). Leaves with enations are misshapen and deeply laciniated, and shoots are variously deformed and sometimes cracked in the basal internodes. Symptoms are erratic and do not recur every year on the same vines.

NATURAL SPREAD
No vector is known. Spread is through propagative material, which perpetuates the disease.

———
Described by Hewitt, 1954.

DETECTION
Strongly symptomatic vines are readily detected in the field, even at a distance, in the early stages of vegetation, because of the bushy compact growth. These vines usually bear outstanding foliar enations that make the disease conspicuous. A search for symptoms in summer is not advisable because the vines resume their normal growth and vigour, and enation-bearing leaves are prematurely shed.

IDENTIFICATION
The only method available is graft transmission to the indicator LN 33 (Martelli *et al.*, 1966). Indexing, however, is highly unsatisfactory. Symptom appearance is slow (it may take up to three years) and occurs at a rate seldom exceeding 20 percent. Symptoms consist of the development of enations on the underside of a few leaves of the indicator (Figure 112). Symptoms should be read when shoots are 15 to 20 cm long, for enations are no longer produced in summer and tend to disappear from the leaves that bear them.

SANITATION
No information is available.

REFERENCES

Graniti, A., Martelli, G.P. & Lamberti, F. 1966. Enation disease of grapevine in Italy. *Proc. Int. Conf. Virus Vectors Perennial Hosts and* Vitis,

1965, p. 293-306. Div. Agric. Sci., Univ. Calif., Davis.

Hewitt, W.B. 1954. Some virus and virus-like diseases of grapevine. *Calif. Dept. Agric. Bull.*, 39: 47-64.

Martelli G.P., Graniti, A., Lamberti, F. & Quacquarelli, A. 1966. Trasmissione per innesto della malattia delle enazioni. *Phytopathol. Mediterr.*, 5: 122-124.

Summary: enation detection

GRAFT TRANSMISSION
Indicators
LN 33 or *Vitis vinifera* cv. Italia
No. plants/test
3-5 rooted cuttings
Inoculum
Single buds, bud sticks
Temperature
Field conditions
Symptoms
Enations and leaf deformation 1-3 years after grafting

FIGURE 110
Severely malformed basal leaves of a European grape
cultivar with outstanding enations

FIGURE 111
Enations on the underside of a leaf, clustered along
the main veins

FIGURE 112
Enations on indicator LN 33
(Photo: U. Prota)

Vein necrosis
G.P. Martelli

CAUSAL AGENT
The causal agent of vein necrosis is unknown.

GEOGRAPHICAL DISTRIBUTION
The disease was first identified in France by Legin and Vuittenez (1973). It is now recorded from most European and Mediterranean countries, where it often has a high level of incidence (Credi, Babini and Canova, 1985; Savino, Boscia and Martelli, 1985), and it is likely to have a much wider distribution.

ALTERNATE HOSTS
No alternate hosts are known.

FIELD SYMPTOMS
There are no field symptoms. The disease is latent in all European grapevine cultivars and in most American rootstock species and hybrids.

NATURAL SPREAD
No vector is known. Spread is through infected propagating material.

DETECTION
No field detection is possible because of the lack of symptoms in affected vines.

IDENTIFICATION
Grafting is done to *Vitis rupestris* x *Vitis berlandieri* 110 R. Symptoms consist of necrosis

of the veinlets on the underside of the leaf blade (Figure 113). The necrotic reactions develop first in the leaves at the base of the shoots and then, as the shoots grow, on the younger leaves. With time, necrotic spots also appear on the upper side of the leaf blade. Severe strains may induce necrosis of tendrils and dieback of green shoots. An almost complete cessation of growth ensues, and the indicator may die. In the field or greenhouse, symptoms appear six to eight weeks after inoculation, but mild strains may induce positive reactions the year after grafting. The symptoms are clearly shown, persist throughout the vegetative season and are readily recorded. Reading of symptoms should be made eight to ten weeks after inoculation.

SANITATION
Heat treatment at 38°C for no less than 60 days, removal and rooting under mist of shoot tips 0.5 cm long eliminates the disease from about 65 percent of the explants (Savino, Boscia and Martelli, 1985).

REFERENCES

Credi, R., Babini, A.R. & Canova, A. 1985. Occurrence of grapevine vein necrosis in the Emilia-Romagna region (northern Italy). *Phytopathol. Mediterr.*, 24: 17-23.

Legin, R. & Vuittenez, A. 1973. Comparaison des symptômes et transmission par greffage d'une mosaïque nervaire de *Vitis vinifera*, de la marbrure

---
Described by Legin and Vuittenez, 1973.

de *V. rupestris* et d'une affection nécrotique de l'hybride *Rup-Ber* 110 R. *Riv. Patol. Veg.*, 9(suppl.): 57-63.

Savino, V., Boscia, D. & Martelli, G.P. 1985. Incidence of some graft-transmissible virus-like diseases of grapevine in visually selected and heat-treated stocks from southern Italy. *Phytopathol. Mediterr.*, 24: 204-207.

Summary: vein necrosis detection

GRAFT TRANSMISSION
Indicator
American *Vitis* hybrid 110 R
No. plants/test
3-5 rooted cuttings
Inoculum
Wood chips, single buds, bud sticks, shoot tips
Temperature
26°C (green grafting)
Symptoms
Necrosis of the veinlets, stunting and necrosis of the shoot tips

FIGURE 113
Necrosis of the veinlets typically induced by vein necrosis in
the indicator 110 R

VIRUS-LIKE DISEASES
Vein mosaic and summer mottle

G.P. Martelli

These diseases are caused by undetermined but possibly distinct aetiological agents (Woodham and Krake, 1983). They are treated together here for practical purposes because of the similarity of their symptoms.

CAUSAL AGENT
The causal agent(s) are unknown. A viroid aetiology was suggested for summer mottle because of the influence of high temperatures on symptom expression (Woodham and Krake, 1983).

GEOGRAPHICAL DISTRIBUTION
Vein mosaic, first identified in France (Legin and Vuittenez, 1973), is widespread throughout Europe and the Mediterranean. Summer mottle was recorded from Australia (Krake and Woodham, 1978), but it is not known whether or not it is restricted to that country.

ALTERNATE HOSTS
No alternate hosts are known.

FIELD SYMPTOMS
Vein mosaic is a semi-latent disease, for it does not show up consistently in affected vines, regardless of whether they are European scion varieties or American rootstocks. The characterizing symptoms consist of pale green discolorations of the tissues adjacent to the main

veins, producing a feathering or banding effect (Figures 114 and 115). The size and vigour of the vines may be adversely affected. Leaf symptoms induced by summer mottle are very similar. The symptoms of both diseases appear in summer and persist through the autumn; sometimes they may be more obvious in greenhouse-grown vines than in the field.

NATURAL SPREAD
No vector is known. Spread is through infected propagating material.

DETECTION
Field detection of vein mosaic is not always possible because of the semi-latency of the disease. Also, when shown, symptoms may escape observation because of their mildness and irregular distribution in the vines.

IDENTIFICATION
The main indicator for vein mosaic is *Vitis riparia* Gloire de Montpellier, which reacts with chlorotic blotches and green mosaic along the veins (Figure 116), malformations of the leaf blade and occasional necrosis. Under greenhouse conditions, symptoms appear four to five weeks after inoculation, reaching full expression in a couple of months. Symptomatological responses may be slower in the field, but they usually show on the vegetation in the first year after grafting. LN 33 is another indicator for vein mosaic, reacting with pale green to yellowish vein banding.

Described by Legin and Vuittenez, 1973; Krake and Woodham, 1978.

Summer mottle does not induce symptoms in *V. riparia* or LN 33 (Woodham and Krake, 1983). Good indicators are European cultivars Sideritis, Cabernet franc and Mission, which exhibit a typical pale green to yellowish vein feathering and vein banding. Symptoms develop both in greenhouse and field in the first year's vegetation of inoculated vines.

SANITATION

Freedom from summer mottle has been obtained by regenerating vines from fragmented shoot apices grown *in vitro* at temperatures between 20°C (night) and 27°C (day) or at a continuous 35°C (Barlass *et al.*, 1982).

Summary: vein mosaic and summer mottle detection

GRAFT TRANSMISSION
Indicators
Vitis riparia (vein mosaic), *Vitis vinifera* cv. Sideritis, Cabernet franc (summer mottle)
No. plants/test
3-5 rooted cuttings
Inoculum
Wood chips, single buds, bud sticks, shoot tips
Temperature
22°C (green grafting) or field conditions
Symptoms
Chlorotic blotches and green mosaic along the veins, leaf deformation in 4-6 weeks

REFERENCES

Barlass, M., Skene, K.G.M., Woodham, R.C. & Krake, L.R. 1982. Regeneration of virus-free grapevines using *in vitro* apical culture. *Ann. Appl. Biol.*, 101: 291-295.

Krake, L.R. & Woodham, R.C. 1978. Grapevine vein mottle, a new graft transmissible disease. *Vitis*, 17: 266-270.

Legin, R. & Vuittenez, A. 1973. Comparaison des symptômes et transmission par greffage d'une mosaïque nervaire de *Vitis vinifera*, de la marbrure de *V. rupestris* et d'une affection nécrotique des nervures de l'hybride *Rup-ber* 110 R. *Riv. Patol. Veg.*, 9(suppl.): 57-63.

Woodham, R.C. & Krake, L.R. 1983. A comparison of grapevine summer mottle and vein mosaic diseases. *Vitis*, 22: 247-252.

FIGURE 115
Intense chlorotic vein banding induced by
vein mosaic disease

FIGURE 114
Chlorotic vein feathering induced by
vein mosaic disease

FIGURE 116
Vein mosaic symptoms in the indicator *V. riparia*

Asteroid mosaic

G.P. Martelli

CAUSAL AGENT
The causal agent is unknown.

GEOGRAPHICAL DISTRIBUTION
Asteroid mosaic was first identified by Hewitt and Goheen (1959) in California and more recently in Greece (P.E. Kyriakopoulou, personal communication).

ALTERNATE HOSTS
No alternate host is known.

FIELD SYMPTOMS
Translucent spots with a star-like shape and fused lateral veins centred between primary veins in the blade are present on the leaves. Leaves are also asymmetrical and crinkled (Figure 117). Diseased vines are weak and bear little or no fruit.

NATURAL SPREAD
No vector is known. Spread is through infected propagating material.

DETECTION
Infected vines are symptomatic and can be identified in the field.

IDENTIFICATION
Four to eight weeks after graft inoculation to *Vitis rupestris* St George, narrow cream-yellow

bands develop on the main veins of leaves (Figure 118). Leaves are distorted and small (Hewitt *et al.*, 1962).

SANITATION
No information is available.

REFERENCES

Hewitt, W.B. & Goheen, A.C. 1959. Asteroid mosaic of grapevines in California. *Phytopathology*, 49: 541.

Hewitt, W.B., Goheen, A.C., Raski, D.J. & Gooding, G.V. 1962. Studies on virus and virus-like diseases of the grapevine in California. *Vitis*, 3: 57-83.

Described by Hewitt and Goheen, 1959.

Summary: asteroid mosaic detection

GRAFT TRANSMISSION
Indicator
Vitis rupestris St George
No. plants/test
3-5 rooted cuttings
Inoculum
Wood chips, single buds, bud sticks
Temperature
Field conditions
Symptoms
Cream-yellow bands along the veins of the indicator and leaf deformation in 6-8 weeks

FIGURE 117
Foliar symptoms of asteroid mosaic
(Photo: U. Prota)

FIGURE 118
Foliar reactions of *V. rupestris* to asteroid mosaic infection

DISEASES INDUCED BY PHLOEM- AND XYLEM-LIMITED PROKARYOTES

Flavescence dorée

A. Caudwell and G.P. Martelli

CAUSAL AGENT

Flavescence dorée (FD) is caused by a mycoplasma-like organism (MLO).

GEOGRAPHICAL DISTRIBUTION

The disease was first recorded in 1954 in the southwest of France (Armagnac), from where it spread to other southern viticultural districts of continental France, Corsica and northern Italy (Caudwell and Larrue, 1986). Flavescence dorée was apparently introduced into Europe from the Great Lakes area of the United States, the home of its vector, where a similar disease occurs in *Vitis vinifera* (Caudwell and Dalmasso, 1985; Pearson *et al.*, 1985).

ALTERNATE HOSTS

No natural host other than the grapevine is known.

FIELD SYMPTOMS

Symptoms usually appear in late spring. Growth can be reduced and the internodes are shortened. The leaves have downward rolled margins, and the shoots may exhibit a drooping condition because of irregular maturation of the wood (Figure 119) and lack of phloem fibres. As the season progresses, the severity of symptoms increases: rolling of the leaves becomes more intense and the blades are discoloured, turning

yellow in white-berried cultivars (Figures 120 and 121) or red in red-berried cultivars (Figures 122 and 123). Necrosis along the main veins may develop in autumn. The internodes may show black pustules (Figure 124) and sometimes longitudinal splitting of the bark. The crop is much reduced. Inflorescences may dessicate when symptoms first appear. If bunches are formed, the berries wither and dry up or drop at the slightest shaking of the vine. Some affected vines die, usually in the year following infection. The survivors recover naturally and do not show symptoms unless they are infected anew. Wild American *Vitis* species are infected without showing symptoms.

NATURAL SPREAD

The disease is transmitted in nature by *Scaphoideus titanus* (= *Scaphoideus littoralis*), a strictly ampelophagous leafhopper species (Figures 125 and 126) that in Europe thrives in southern France, Corsica, southern Switzerland and northern Italy. This leafhopper lays eggs in the bark of two-year-old grape wood and has five larval stages and a single generation per year. Local spread is effectively accomplished by the vector, whereas long-distance dissemination is through infected propagative material.

DETECTION

Detection is based on observation of field symptoms in summer or autumn.

Described by Caudwell, 1964.

IDENTIFICATION

Indexing by graft transmission

The most sensitive indicator is the hybrid variety Baco 22A (Noah x Folle blanche). Chip-bud and cleft grafting are both suitable for inoculation. Grafted indicators show symptoms within two to three months. The symptoms are the same as those shown by naturally infected plants in the field, i.e. stunting and downward rolling and yellowing of the leaves. Aramon, Chardonnay, Sangiovese and Alicante Bouchet are other sensitive cultivars that can also be used as indicators.

Transmission by vectors

Adults and larvae of *S. titanus* readily transmit the disease agent to broad bean (*Vicia faba*) cv. Arla under greenhouse conditions. Colonies of *S. titanus* can be obtained by placing egg-bearing two-year-old grape pruning wood kept moist in a cage around vegetating vines (Caudwell and Larrue, 1977). Leafhoppers (fourth or fifth stage instars or adults) are placed on infected vine shoots for three to four weeks, then transferred for one to two weeks on to young (4 to 5 cm tall) broad bean seedlings. Inoculated broad beans are removed from the cage and grown in the greenhouse with 16-hour artificial illumination. Symptoms consisting of reduced growth and yellowing and upward rolling of the leaves accompanied by progressively severe flower abortion (Figure 127) develop at the top of the plant within two to three weeks.

Euscelidius variegatus, another leafhopper species, can be used to transmit the causal agent of the disease (FD-MLO) from broad bean to broad bean as well as to other plant species such as *Vicia, Pisum, Chrysanthemum* and *Lupinus* species and *Vinca rosea*. Since *E. variegatus* does not thrive on *Vitis* species, its colonies are established and maintained on maize (*Zea mais*) or broad bean under controlled conditions.

Serology

To minimize non-specific reactions in serological identification of FD-MLO, antisera to *E. variegatus* FD can be used against extracts of *V. faba* FD. Conversely, antisera to *V. faba* FD can be used against extracts of leafhopper FD, i.e. extracts from infected *E. variegatus* reared artificially or viruliferous *S. titanus* collected from diseased vines in the field.

ISEM. Freshly carbonated grids are coated with a 10 µg/ml solution of protein A in 0.1 M phosphate buffer, ph 7.2. After rinsing in the same buffer, the grids are sensitized by floating on a drop of undiluted antiserum to *E. variegatus* FD, are again rinsed in buffer and are floated for 15 minutes on a drop of infected broad bean leaf extract containing 0.05 percent Tween 20. After thorough rinsing with phosphate buffer containing 0.05 percent Tween 20, the grids are fixed with 1 percent glutaraldehyde, rinsed with distilled water and negatively stained with 2 percent ammonium molybdate. In infected samples FD-MLOs appear as vesicles. Similar results are obtained by using an antiserum to *V. faba* FD and extracts from infected leafhoppers.

ELISA. Individual leafhoppers are crushed in 0.5 ml of PBS buffer, and 0.1 ml of the extract is placed in a plate well. Anti-*V. faba* FD rabbit immunoglobulins are added. The conjugate is an anti-rabbit goat IgG. If infected leaf extracts are employed for ELISA, the IgGs used are rabbit anti-leafhopper FD. Both polyclonal and monoclonal antibodies can be used satisfactorily in ELISA (Boudon-Padieu and Larrue, 1986; Boudon-Padieu, Larrue and Caudwell, 1989; Schwartz *et al.*, 1989).

SANITATION

Spreading of the disease occurs through infected propagation material and the vector. Symptomless grape budwood can host both the eggs of the leafhopper vector and the disease agent. Thus nurseries must be established in, and propagating material must be collected from, areas that are free from the disease or particularly well protected against the vector. The vector is controlled by: eliminating eggs by burning pruning wood, treating before bud burst with parathion-activated oils (1 500 ml of a suspension of 3 percent parathion and 78 percent paraffin oil per 100 litres of water); or one or two chemical applications against instars 30 and 45 days after the first hatching, followed by another treatment against adults in August.

FD-MLOs may be eliminated from infected wood by treating dormant canes with water at 45°C for 3 hours or at 50°C for 45 minutes (Caudwell *et al.*, 1990).

REFERENCES

Boudon-Padieu, E. & Larrue, J. 1986. Diagnostic rapide de la flavescence dorée de la vigne par le test ELISA sur cicadelle vectrice. Application à des populations naturelles de *Scaphoideus littoralis* Ball. Confirmation de la présence de la flavescence dorée dans les Bouches-du-Rhône. *Prog. Agric. Vitic.*, 103: 524-526.

Boudon-Padieu, E., Larrue, J. & Caudwell, A. 1989. ELISA and dot-blot detection of flavescence dorée-MLO in individual leafhopper vectors during latency and inoculative state. *Curr. Microbiol.*, 19: 357-364.

Caudwell, A. 1964. Identification d'une nouvelle maladie à virus de la vigne, la flavescence dorée. Etude des phénomènes de localisation des symptômes et de rétablissement. *Ann. Epiphytol.*, 15(hors sér. I). 197 pp.

Caudwell, A. & Dalmasso, A. 1985. Epidemiology and vectors of grapevine viruses and yellows diseases. *Phytopathol. Mediterr.*, 24: 170-176.

Caudwell, A. & Larrue, J. 1977. La production de cicadelles saines et infectieuses pour les épreuves d'infectivité chez les jaunisses à mollicutes des végétaux. L'élevage de *Euscelis plebeja* KBM et la ponte sur mousse de polyuréthane. *Ann. Zool. Ecol. Anim.*, 9: 443-456.

Caudwell, A. & Larrue, J. 1986. La flavescence dorée dans le midi de la France et dans le Bas-Rhône. *Prog. Agric. Vitic.*, 103: 517-523.

Caudwell, A., Larrue, J., Volos, C. & Grenan, S. 1990. Hot water treatment against flavescence dorée on dormant wood. *Proc. 10th Meet. ICVG*, Volos, Greece, 1990, p. 336-343.

Pearson, R.C., Pool, R.M., Gonsalves, D. & Goffinet, M.C. 1985. Occurrence of flavescence dorée-like symptoms on "White Riesling" grapevines in New York, USA. *Phytopathol. Mediterr.*, 24: 82-87.

Schwartz, Y., Boudon-Padieu, E., Grange, J., Meignoz, R. & Caudwell, A. 1989. Obtention d'anticorps monoclonaux spécifiques de l'agent pathogène de type mycoplasme (MLO) de la flavescence dorée de la vigne. *Ann. Inst. Pasteur Res. Microbiol.*, 140: 311-324.

Summary: flavescence dorée detection

GRAFT TRANSMISSION
Indicators
Hybrid Baco 22A; *Vitis vinifera* cvs Chardonnay, Aramon
No. plants/test
3-5 rooted cuttings
Inoculum
Wood chips, single buds, bud sticks
Temperature
Field conditions
Symptoms
Stunting, leaf yellowing and necrosis (white-berried varieties) or leaf reddening and necrosis (red-berried varieties) 2-3 months or more after inoculation

OTHER TESTS
Serology (ELISA)
Molecular hybridization

FIGURE 119
Irregular wood ripening in a shoot of a vine
affected by flavescence dorée

FIGURE 122
Intense reddening caused by flavescence dorée
infection in cv. Pinot noir

FIGURE 120
Vine of Baco 22A with stunting and yellowing caused
by flavescence dorée infection (left) next to a healthy
vine (right)
(Photo: G. Granata)

FIGURE 123
Reduced vigour, stunting and leaf reddening of
cv. Aramon caused by flavescence dorée

FIGURE 121
Close-up of Baco 22A leaves with symptoms of
flavescence dorée. Note shelling and shrivelling of
the cluster

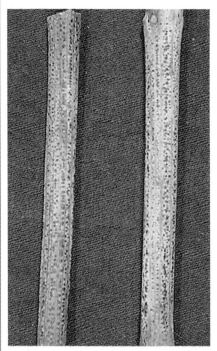

FIGURE 124
Black pustules on the canes of a vine
infected by flavescence dorée
(Photo: G. Granata)

FIGURE 126
Adult of *Scaphoideus titanus*
(Photo: A. Brun)

FIGURE 125
Instar of the flavescence dorée vector
Scaphoideus titanus
(Photo: A. Brun)

FIGURE 127
Broad bean plant artificially infected with the agent of
flavescence dorée (right) next to a healthy plant (left)

DISEASES INDUCED BY PHLOEM- AND
XYLEM-LIMITED PROKARYOTES
Grapevine yellows

G.P. Martelli and A. Caudwell

CAUSAL AGENT

The causal agent of grapevine yellows (bois noir, Vergilbungkrankheit, etc.) is unknown, but it is thought to be a mycoplasma-like organism (MLO). Whether one or more different MLOs are involved in the aetiology of the disease is not known.

GEOGRAPHICAL DISTRIBUTION

Syndromes comparable to bois noir, one of the grapevine yellows originally reported from France (Caudwell, 1961), have been recorded from many European and Mediterranean countries (Germany, Switzerland, southern Italy, Romania, Yugoslavia, Greece, Israel). Similar diseases occur also in Chile, Argentina and Australia (see Caudwell, 1988). Whether and to what extent these diseases are related to one another has not been established.

ALTERNATE HOSTS

No alternate hosts are known, although they are likely to exist in native flora (weeds and shrubs).

FIELD SYMPTOMS

Symptoms are practically the same as those of flavescence dorée, i.e. leaf rolling, yellowing or reddening of the leaves, necrosis along the veins, incomplete wood ripening, withering of berries and drying up of bunches (Figures 128 to 131).

_____
Described by Caudwell, 1961.

NATURAL SPREAD

Visual evidence strongly indicates that the disease spreads naturally in the field. Vectors are not yet known but are likely to be leafhoppers.

DETECTION

Infected vines are readily identified because of the symptoms shown, especially in summer, which is the best time for field surveys.

IDENTIFICATION

Although cv. Chardonnay is a sensitive indicator for certain of these diseases (e.g. bois noir), it may not be totally dependable for others. Vines can be infected by grafting but not by *Scaphoideus titanus*. The symptomatological reactions of cv. Chardonnay tally with field symptoms. Serological tests are not available.

SANITATION

No information is available.

REFERENCES

Caudwell, A. 1961. Etude sur la maladie du bois noir de la vigne: ses rapports avec la flavescence dorée. *Ann. Epiphytol.*, 12: 241-262.

Caudwell, A. 1988. Bois noir and Vergilbung-krankheit. Other grapevine yellows. *In* R.C. Pearson and A.C. Goheen, eds, *Compendium of grape diseases*, p. 46-47. St Paul, MN, USA, Am. Phytopathol. Soc.

Summary: yellows detection

GRAFT TRANSMISSION
Indicator
Vitis vinifera cv. Chardonnay
No. plants/test
3-5 rooted cuttings
Inoculum
Wood chips, single buds, bud sticks
Temperature
Field conditions
Symptoms
Yellowing or reddening of the leaves and rolling of the
blades followed by necrosis of the veins, usually within
the first year after inoculation

FIGURE 128
View of a white-berried grapevine cultivar
affected by bois noir
(Photo: G. Granata)

FIGURE 130
Sectorial reddening typical of bois noir
infection in a red-berried cultivar
(Photo: G. Granata)

FIGURE 129
Severe yellowing of the leaf blade and
necrosis of the main vein induced by
bois noir infection in a white-berried
cultivar
(Photo: G. Granata)

FIGURE 131
Leaf yellowing and drying up of bunches
in a vine affected by bois noir
(Photo: G. Granata)

DISEASES INDUCED BY PHLOEM- AND
XYLEM-LIMITED PROKARYOTES
Pierce's disease
D.A. Golino

CAUSAL AGENT

Xylella fastidiosa, a gram-negative aerobic bacterium found only in the xylem tissue of infected plants (Wells *et al.*, 1987; Davis, Whitcomb and Gillaspie, 1981), has recently been named as the causal agent of Pierce's disease (PD).

GEOGRAPHICAL DISTRIBUTION

PD was first described in 1892 in southern California by N.B. Pierce, after whom it was named. PD has been held responsible for the destruction of the once-extensive grape plantings in that area. It is widespread throughout areas of the Western Hemisphere with mild winter temperatures, including all of the southern United States from California to Florida. It limits the cultivation of *Vitis vinifera* in many of these areas. Since the development of techniques for cultivation and serological detection of the pathogen have facilitated identification of the causal agent, Chile, Costa Rica, Mexico and Venezuela have been added to the list of countries in which PD occurs. Reports of infection outside the Americas have recently been made (Boubals, 1989).

ALTERNATE HOSTS

Other *Vitis* species are hosts of *X. fastidiosa*; native *Vitis* species are believed to serve as

_____
Described by Hewitt *et al.*, 1942.

reservoirs of infection in much of the southeastern United States (Hopkins, 1988). In addition, the pathogen has a wide natural host range that includes both annual and perennial plants of many genera (Raju, Goheen and Frazier, 1983). Many plant hosts show either no symptoms or mild symptoms. *Xylella fastidiosa* causes several diseases, including almond leaf scorch (Figure 132), alfalfa dwarf and a disease of macadamia in Costa Rica. Other *Xilella* species are known to cause phony peach disease and a scorching disease of numerous tree species. The taxonomic relationship of *X. fastidiosa* to the strains that cause these diseases is as yet undetermined (Hopkins, 1988).

FIELD SYMPTOMS

Obvious PD symptoms are scorched or dry leaves on one or a few canes. Symptoms are usually spread asymmetrically in the canopy until late in the development of the disease. Infected vines often exhibit delayed bud break in the spring.

Early season foliage may show interveinal chlorosis (Figure 133) and develops more slowly than foliage on surrounding healthy vines. In midsummer, as water stress from vascular plugging begins to affect the plant, asymmetrical yellowing of the leaf margins is seen. These discoloured regions become progressively more necrotic until autumn (Figures 134 to 136). Yields are reduced, and clusters are often shrivelled (Figure 137). Leaves fall prematurely, leaving the petiole intact on the vine (Figure 138). This symptom is considered diagnostic for PD. Canes mature unevenly in the autumn, leaving patches of immature green wood (Figures 139 and 140).

Vines deteriorate rapidly after appearance of symptoms. Infected plants grow progressively weaker as symptoms become more pronounced (Figure 141). PD is normally fatal to infected vines. The life expectancy of diseased vines is reported to vary from one to a few years.

The symptoms of PD can be easily confused with other phenomena that cause water stress. In addition, some other diseases such as Eutypa dieback (*Eutypa armeniaca*), oakroot fungus (*Armillariella mellea*) and measles may cause similar symptoms. Nutrient imbalances can also cause chlorosis and scorching.

NATURAL SPREAD

PD is spread by xylem-feeding insects which insert their mouthparts directly into those tissues in which the bacteria are found (Figure 142). All species of sharpshooters (Cicadellidae) and spittle bugs (Cercopidae) that have been tested under experimental conditions are capable of transmitting the bacterium, whereas phloem-feeding leafhoppers, which occasionally probe xylem tissues, do not transmit the disease. This suggests that feeding behaviour is an important component of the relationship between the vectors and *X. fastidiosa* (Purcell, 1989). The bacteria can be readily observed in large numbers with the scanning electron microscope on the surfaces of the feeding apparatus of vector species (Figure 143).

In the western United States spread into vineyards is thought to occur from surrounding vegetation. Vine-to-vine spread is not believed to contribute significantly to diseases in these areas (Purcell, 1974). In contrast, in the southeastern United States diseased grapes may serve as sources of inoculum (Goheen and Hopkins, 1988).

DETECTION

PD is normally detected by the observation of symptoms in the late summer and autumn. Detection is also possible in the dormant season or early spring by experienced workers.

IDENTIFICATION

Symptomatology

PD can be diagnosed with some confidence in European grapes on the basis of symptoms during the late summer and autumn (Figures 134 to 140). Spring symptoms (Figure 133) are more easily confused with other conditions. The pathogen is non-symptomatic in many of its host plants. Absolute identification relies on cultivation of the bacterium on selective media and/or the use of serological techniques.

Cultivation

Xylella fastidiosa cannot grow on commonly used bacteriological media. Specialized media have thus been developed (Hopkins, 1988). A method for isolating and growing the PD bacterium is outlined in Part III. Although the PD3 medium is quite satisfactory, other media may be required for some pathovars. Petioles from symptomatic leaves are ideal for isolation purposes, although other grapevine parts may be acceptable. *Xylella fastidiosa* is slow growing, and colonies may require one to three weeks to develop. Colonies are white and smooth, with complete margins (Davis, Purcell and Thompson, 1978).

Serology

Antisera to *X. fastidiosa* are readily prepared by the use of either whole cells (Davis, Purcell and Thompson, 1978) or sonicated cultured cells (Hopkins, 1988) as antigen. Antisera and type culture of the bacterium are both available from the American Type Culture Collection (ATCC, 12301 Parklawn Drive, Rockville, MD 20852, USA). Purified antibodies can be used either in ELISA or in Outcherlony gel diffusion tests as described in Part III. No serological differences have been observed between the strains of *X. fastidiosa* that cause PD and those that cause elm leaf scorch, phony peach, plum leaf scald and periwinkle wilt (Hopkins, 1988).

Electron microscopy

When fixed and prepared for electron microscopy, *X. fastidiosa* can be seen as rod-shaped bacterial elements 0.25 to 0.5 µm in diameter and 1.0 to 4.0 µm in length (Mollenhauer and Hopkins, 1974). A rippled cell wall is characteristic of the species.

In cross-sections through the xylem elements of infected grapes, bacteria can be seen blocking some vessels, while other vessels remain clear (Figure 144).

Pathogenicity

Inoculation of grapes and other plants is readily accomplished by needle inoculation with the cultured bacterium. A drop of turbid suspension of the bacterium in phosphate-buffered saline is placed on a leaf or in the angle of a petiole, and a sterile needle is inserted through the droplet into the plant tissues. Xylem tension will draw the inoculum into the plant. A syringe filled with the inoculum may also be used. On grapes, PD symptoms develop rapidly. Pathogenicity of the cultured organism is easily lost in culture, so early passages of new isolates should be preserved by storing at -20°C if the pathogenicity of an isolate needs to be determined or maintained.

It is also possible to transmit PD by grafting from infected to healthy vines. The graft must contain the active xylem tissues which harbour the PD agent. This technique can provide a useful diagnostic check if culture media, antisera or electron microscope facilities are not available. In pathogenicity testing one should always keep in mind that avirulent and mild strains of the bacterium are known to occur (Hopkins, 1984).

SANITATION

PD is normally lethal for individual vines. Although treatment with antibiotics will result in a remission of symptoms, the expense of this treatment, the recurrence of symptoms when the treatment is discontinued and environmental concerns about field applications of antibiotics make chemical therapy infeasible. Control strategies have been based largely on eliminating vector species or host-plant inoculum. Knowledge of the identity of the natural vector species and alternative plant hosts in a given region is an essential component of the development of sanitary procedures.

Given the wide host range of the PD pathogen and the large number of sharpshooter species that are vectors, control of the disease in areas where *X. fastidiosa* is established in native vegetation is extremely difficult. In the Central Valley of California, PD is transmitted largely by the green- and red-headed sharpshooters *Draeculacephala minerva* and *Carneocephala fulgida*. Field studies have established that the alternate hosts of these sharpshooters include weeds growing in alfalfa fields and some of the perennial grasses that grow at the margins of the fields. Elimination of the weed species and application of pesticides to neighbouring alfalfa fields have been effective in controlling disease.

The north coast valleys of California have a high incidence of PD foci. Control of alternate hosts of the pathogen and elimination of vector populations are difficult since they originate in riparian areas which are protected as wildlife refuges. Only limited control of vector species is possible with pesticides, since applications to woodlands are both impractical and illegal. Often a border effect is observed, with a high mortality of vines adjacent to native vegetation. Vine-to-vine spread is not believed to be of importance in these locations.

In the Gulf Coastal Plains of the United States, eastern coastal Mexico and the tropical Americas, *X. fastidiosa* is well established in native *Vitis* species and other vegetation. Neither *V. vinifera* nor *Vitis labrusca* survives more than a few years in these areas before becoming infected. Control strategies are limited to planting of PD-resistant varieties of native American genera such as *Muscadinia* or *Euvitis* and the development of resistant hybrids between those species and their more susceptible relatives.

Quarantine regulations that limit the movement of *Vitis* species from the Americas are aimed at preventing the introduction of *X. fastidiosa* from the warm grape-growing regions where PD is common into other regions. Fortunately, dormant cuttings harbouring the pathogen are normally short-lived. In addition, hot water treatment of dormant cuttings (immersion at 45°C for three hours or at 50°C for 20 minutes) will destroy *X. fastidiosa* (Goheen, Nyland and Lowe, 1973).

REFERENCES

Boubals, D. 1989. La maladie de Pierce arrive dans les vignobles d'Europe. *Prog. Agric. Vitic.*, 106: 85-87.

Davis, M.J., Purcell, H.A. & Thompson, S.V. 1978. Pierce's disease of grapevines: isolation of the causal bacterium. *Science*, 199: 75-77.

Davis, M.J., Whitcomb, R.F. & Gillaspie, A.G.M. 1981. Fastidious bacteria of plants and insects (including so-called rickettsia-like bacteria). *In* M.P. Starr, H.O. Stolp, H.G. Truper, A. Balows & H.G. Schelegel, eds, *The prokaryotes: a handbook on habitats, isolation and identification of bacteria*, 2: 2171-2188. Berlin, Springer-Verlag.

Goheen, A.C. & Hopkins, D.L. 1988. Pierce's disease. *In* R.C. Pearson and A.C. Goheen, eds, *Compendium of grape diseases*, p. 44-45. St Paul, MN, USA, Am. Phytopathol. Soc.

Goheen, A.C., Nyland, G. & Lowe, K.S. 1973. Association of a rickettsialike organism with Pierce's disease of grapevines and alfalfa dwarf and heat therapy of the disease in grapevines. *Phytopathology*, 63: 341-345.

Hewitt, W.B., Frazier, N.W., Jacob, H.E. & Freitag, H.J. 1942. *Pierce's disease of grapevines.* Calif. Agric. Exp. Sta. Circ. No. 353. 32 pp.

Hopkins, D.L. 1984. Physiological and pathological characteristics of virulent and avirulent strains of the bacterium that causes Pierce's disease of grapevine. *Phytopathology*, 75: 713-717.

Hopkins, D.L. 1988. *Xylella fastidiosa* and other bacteria of uncertain affiliation. *In* N.W. Schaad, ed., *Laboratory guide for the identification of plant pathogenic bacteria*, 2nd ed., p. 95-103. St Paul, MN, USA, Am. Phytopathol. Soc.

Mollenhauer, H.H. & Hopkins, D.L. 1974. Ultrastructural study of Pierce's disease bacterium in grape xylem tissue. *J. Bacteriol.*, 119: 612-618.

Purcell, A.H. 1974. Spatial patterns of Pierce's disease in the Napa Valley. *Am. J. Enol. Vitic.*, 25: 162-167.

Purcell, A.H. 1989. Homopteran transmission of xylem-inhabiting bacteria. *Adv. Dis. Vector Res.*, 6: 243-266.

Raju, B.C., Goheen, A.C. & Frazier, N.W. 1983. Occurrence of Pierce's disease bacterium in plants and vectors in California. *Phytopathology*, 73: 1309-1313.

Wells, J.M., Raju, B.C., Hung, H.Y., Weisburg, W.G., Mandelco-Paul, L. & Brenner, D.G. 1987. *Xylella fastidiosa* gen. nov., sp. nov.: gram negative, xylem-limited, fastidious plant bacteria related to *Xanthomonas* spp. *Int. J. Syst. Bacteriol.*, 37: 136-143.

Summary: Pierce's disease detection

GRAFT TRANSMISSION

Indicators
Several *Vitis vinifera* cultivars (Chardonnay, Merlot)

No. plants/test
3-5 rooted cuttings

Inoculum
Wood chips, single buds, bud sticks

Temperature
Field conditions

Symptoms
Scorching, yellowing or reddening of the leaves within the first year after inoculation

OTHER TESTS
Serology (immunodiffusion, ELISA)

FIGURE 132
Symptoms of leaf scorching in almond infected
with *Xylella fastidiosa*, the Pierce's disease agent

FIGURE 135
Typical autumn symptoms of Pierce's disease
in cv. Chardonnay in California
(Photo: A. Yen)

FIGURE 133
Chardonnay with Pierce's disease. A leaf
in early spring exhibits characteristic
interveinal chlorosis
(Photo: P. Goodwin)

FIGURE 136
Close-up of autumn symptoms in cv. Chardonnay

FIGURE 134
Marginal scorching of cv. Merlot leaves caused by
Pierce's disease in autumn

FIGURE 137
Shrivelling of a bunch in a vine
affected by Pierce's disease. Note
the green islands of bark resulting
from uneven maturation of the canes
(Photo: A. Yen)

FIGURE 139
Marginal leaf scorching and uneven wood ripening
of a cv. Chardonnay shoot affected by Pierce's disease

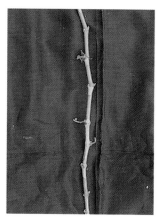

FIGURE 140
Green patches of immature
wood alternating with brown
mature wood in a cv. Chardonnay
cane affected by Pierce's disease

FIGURE 138
Petioles persist after abscission of leaves on canes of
a vine infected with Pierce's disease
(Photo: H. Andris)

FIGURE 141
Final stage of a vine infected by Pierce's disease

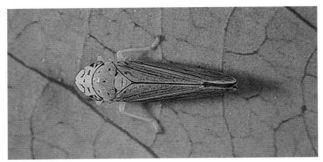

FIGURE 142
The blue-green leafhopper *Graphocephala atropunctata*, one of the
many sharpshooter species known to transmit *Xylella fastidiosa*
(Photo: J. Clark)

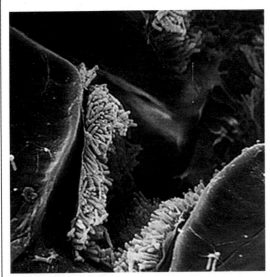

FIGURE 143
Scanning electron micrograph of a mat of *Xylella
fastidiosa* lining the aperture of the feeding stylet of a
sharpshooter vector
(Photo: M.G. Kinsey)

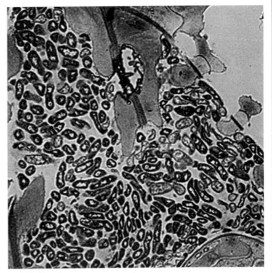

FIGURE 144
Electron micrograph of a cross-section through a
xylem element of a *Xylella fastidiosa*-infected vine.
The neighbouring vessel is free of bacteria
(Photo: A. Purcell)

Part II

Facilities and techniques for identification of diseases and their agents by biological methods

Facilities for growing indicator plants

C.N. Roistacher

THE GREENHOUSE OR PLANT LABORATORY

A greenhouse, or a controlled-environment structure, is necessary for the production of index plants and for indexing (chip-budding prior to transplanting in the field and green grafting). This structure need not be expensive or elaborate. It should provide light, heat and cooling and be sufficiently well constructed to prevent insect intrusion. An entrance with two doors and a darkened vestibule between is desirable as a preventive measure against insect invasion. The greenhouse covering can be of glass or plastic. Excellent plants can be grown in a simply designed and inexpensive wooden structure covered with heavy fibreglass and containing a good system for heating and cooling. Modern structures are now made with extruded aluminium framing.

A sketch of three general greenhouse structures is shown in Figure 145. A number of greenhouse and screenhouse structures in use worldwide are shown in Figures 146 to 156. In areas where hailstorms are frequent, glass coverings should be avoided, but if used they should be protected with wire mesh. Corrugated fibreglass rather than glass is recommended where hail is a problem, and in many respects is preferable since it may be less expensive, does not break and may be easier to construct and maintain.

The size of the greenhouse will depend upon the amount of indexing and research to be carried out. It may be convenient to have two compartments: a cool room for indexing graft-transmissible pathogens which are best expressed in plants grown under relatively cool temperatures (22 to 24°C); and a warm room (up to 34 to 35°C) primarily used for preconditioning vines prior to thermotherapy.

Benches

Benches can be made of wood, concrete, wire mesh, plastic or any satisfactory container-supporting system (Figures 157 to 163). If wooden benches are used, it is advisable to paint or spray them with 2 percent copper naphthenate solution (Roistacher and Baker, 1954), which acts as both a wood preservative and a disinfectant (Figure 164). Wood benches can be placed on concrete blocks or on a metal frame or other foundation at a height of not less than 80 cm from the ground.

Flooring

Flooring can be of concrete with provisions for drainage. However, gravel flooring with a concrete walk is recommended. Gravel of 1 to 2 cm diameter should be spread over the ground about 8 to 10 cm thick. This provides good drainage and aids in maintaining sanitation. The greenhouses should be constructed on a well-drained soil base. If this is not possible, supplementary underfloor drainage tiling should be provided prior to construction.

Containers

Grapevines may be grown in rigid plastic or clay containers of any suitable size. Plastic disposable containers (Figure 165) are useful for transplanting rooted indicators immediately after chip-bud grafting, prior to transfer to the field. For prolonged maintenance of vines under greenhouse or screenhouse conditions, large (25 to 35 cm in diameter and 25 to 35 cm deep) cylindrical or tapered plastic or clay pots (Figure 166) can be used. Drawbacks of clay pots are that they accumulate salts, are heavier, are subject to breakage and must be soaked in water and washed after each use.

Temperature control

It is important to have a recording thermograph in each room. These should be periodically calibrated against two thermometers for accuracy. At the end of each week, when charts are changed, the maximum and minimum temperatures should be recorded in a special book. This provides a record for research and is also a means of noting any abnormal changes, which give warning of heating or cooling unit failure or breakdown.

Supplemental lighting

During the grapevine vegetating season (April to October) no supplemental lighting is required. It is necessary, however, for growing herbaceous indicators in winter months. In the temperate zone (roughly from 30 to 45° N latitude) the addition of 4 to 5 hours of 2 000 to 2 500 lux at the plant level from October to April is useful. Light sources may be fluorescent tubes or incandescent bulbs (e.g. Philips HLRG 400 W) (Figure 167).

Heating

Heating can be provided by gas heaters with fans, by steam heating using radiators or by steam pipes placed along the sides of the structure. Heat may also be distributed from gas heaters using supplementary fans blowing the heat through perforated plastic tubes. Most gas heaters are placed inside the structure. However, ethylene released by faulty heaters can be very damaging to plants. If feasible, gas heaters should be placed outside (Figures 168 and 169) rather than inside (Figure 170) the structure, and the warm air circulated by a fan inside the greenhouse, preferably by forcing the air through large-diameter perforated plastic tubes (Figures 170 to 172). The construction of a greenhouse or screenhouse is best carried out through local builders, with design and facilities suggested by those in charge of indexing.

Cooling

There are three general methods for cooling a greenhouse: introducing air from the outside when the temperature is cooler than that inside the structure; use of evaporative coolers if the relative humidity is low enough to make such cooling effective; and refrigeration. There may be other innovative methods, such as the double-layered plastic bubble which acts both as an insulator and as a sandwich through which cool (or warm) air can be forced. Combinations of any of these methods may be used for economy and efficiency depending upon local conditions.

Air cooling. The simplest and most economical means of cooling a greenhouse is by bringing in outside air to replace the warm air within. This is best accomplished by using fans (Figure 173) and thermostatic control (Figure 174). When the temperature rises the thermostat is activated, the fans turn on and the cooler air is drawn through the greenhouse. A greenhouse designed to utilize the cooling ability of the outside air will save much expensive energy and wear on cooling equipment. Air brought in from outside must be screened or filtered to prevent introduction of insects. Figure 175 shows an air filtering device containing both 32-mesh plastic screen and glass wool filters. This system also has charcoal trays to filter out air pollutants.

The thermostatic controls shown in Figure 174 are designed to control both heating and cooling. As the temperature rises inside the greenhouse the thermostat will activate the fan (Figure 173), thus bringing outside air into and through the house and forcing the warm air outside. When the temperature increases further the thermostat switches on the water pump, which sends water to the cooling cell to begin evaporation cooling (see Figure 178).

Another way of using outside air to cool a greenhouse is to have vents at the peak of the roof. Vents may be activated mechanically, by hand or by a thermostatically controlled motor. When the vents are opened, they permit the warmed inside air to rise and bring in the cooler outside air through filtered vents at the lower sides of the structure. Many problems are associated with this method of air cooling, and though it is present in many older installations, it is generally not recommended for a plant laboratory greenhouse.

Evaporator coolers. Evaporator coolers are recommended for most greenhouses when humidity during summer months is low. An engineering study should be carried out to calculate the cooling ability of evaporator coolers where humidity is moderate or high during the warm months. Evaporator coolers may prove uneconomical and unsound if the relative humidity is too high. However, in some areas evaporator coolers can be combined with refrigeration for efficient cooling.

Equipment and methods for cooling by evaporation are shown in Figures 176 to 179. Figure 176 shows a standard commercial evaporator cooler available in most countries where humidity is low and where homes and buildings are cooled by this means. Figure 177 shows the inside of this cooler with the panel removed to expose the squirrel-cage fan, water

reservoir at the bottom, water pump, water dripping down from the outlet at the top and pads made of wood fibre or glass wool housed inside the panel door. Such units should be carefully serviced each year by cleaning, painting and changing the cooling pads. A standby cooler should be available for emergency replacement, as well as spare water pumps, fan belts and drive motor.

A more efficient apparatus for cooling is shown in Figures 178 and 179. Figure 178 shows cooling cells consisting of rectangular units of specially treated cardboard placed together to form a solid block. Water is pumped from a reservoir tank shown in the lower left of Figure 178 to a trough above the cell. The water then drips down by gravity over the cardboard cells. The outside air is forced through the moistened cells by the diminished pressure induced by the fans located at the opposite end of the greenhouse (Figure 173). The operation of the fans and water pump is controlled by thermostat. Figure 179 illustrates a greenhouse at Lake Alfred, Florida with this cooling system but without an insect filter screen. The cooling cells occupy the full length of the outside wall of the greenhouse.

Refrigeration. Refrigeration can be used to supplement evaporator coolers where the relative humidity is too high during the warmer months, where extra cooling capacity is needed as a supplement for the plants in a cool indexing room or for cooling small individual rooms. Small plastic chambers can be built inside a large greenhouse to give areas of controlled cooling using refrigeration. Figure 180 shows refrigeration units used to cool a grape-indexing facility in South Africa, where electrical energy is relatively inexpensive. Refrigeration is recommended for smaller greenhouses for those compartments to be held at cooler temperatures. These units should be designed to be easily removable for repair and replacement, and a spare unit should be held in reserve to replace any that may need repair.

SOIL MIX FOR PLANT GROWTH

Plants for indexing, regardless of whether they are herbaceous or woody, should be of the highest quality. Therefore, the soil mixture with its balanced supply of micro- and macronutrients is of prime importance. The University of California (UC) system for producing healthy container-grown plants was developed by Baker and colleagues (Baker, 1957), based on the John Innes system of soil mixes developed in England. The system was later modified (Nauer, Roistacher and Labanauskas, 1967, 1968) for growing citrus by the addition of micronutrients to the artificial mixture. This mixture is also suitable for growing grapevines.

Ingredients

The basic soil mixture consists of 50 percent Canadian peat moss and 50 percent fine sand, with macro- and micronutrients added to the mix (Figure 181). Although Canadian peat moss is recommended as the prime ingredient because of its superior nutrient retention and chelating ability, other peats from Europe (e.g. Finland, Poland, Germany, former USSR) can be used after appropriate testing. Alternative ingredients such as wood shavings complemented with extra nitrogen, sphagnum mosses, perlite or vermiculite can be tried whenever necessary.

It is recommended that a fine sand or silt, with a particle size ranging from 0.05 to 0.5 mm, be used. Beach sand should be avoided. Fine sand can be found in rivers, in wind-blown deposits or as the fine silt separated out as waste material from a sand and gravel processing pit. A quick and simple test for determining the presence of clay in a sand source is to shake a sample of the test soil in a jar with water. If the sand settles fairly rapidly and the water remains relatively clear, it is satisfactory. If clay is present, the water will have a muddy appearance, and that source of sand should preferably not be used. The sand should be inert and preferably siliceous. Calcareous and limestone sands should be avoided since they may affect the pH. If a high-grade silicate sand is not available, consideration should be given to a substitute mix of peat, vermiculite and perlite in a ratio of 2:1:1 or 1:1:1. The objective is to obtain an artificial mixture that is consistently reproducible, will absorb and release macro- and micronutrients and will maintain pH of the drainage water at 5.5 to 6.5.

The ingredients can be mixed together with a shovel on a flat concrete surface. However, a small or medium-sized electric or gasoline-powered concrete mixer is the preferred mixing device. In this case the procedure is as follows:

- A specific number of uniform, standard shovel-scoops of soil, peat and wood shavings, or other substitutes for part of the peat moss, are counted and shovelled into the apron of the concrete mixer.
- A weighed quantity of macronutrients, i.e. phosphate, calcium and magnesium, is sprinkled on top of the unmixed ingredients in the apron.
- The soil ingredients plus macronutrients are then dumped into the concrete mixer and thoroughly tumbled.
- The micronutrients, pre-weighed and mixed together in a package, are first dissolved in a container of water and then poured into the turning mixer.
- A small quantity of water can be added to the soil while the mixer is turning to bring the soil mixture to a friable, moist level. This may be necessary if the soil or peat is too dry.

• After about 20 minutes of tumbling and mixing, the soil is emptied from the mixer into a trailer fitted on the bottom with 20-mm galvanized steam pipes with 5-mm holes located at the bottom of the pipes, spaced 15 to 20 cm apart. The pipes are spaced 15 cm apart (Figure 182). The trailer top is covered with a cloth tarpaulin. Containers and flats may be placed on top of the soil in the trailer before covering, or steamed separately (Figures 183 and 184).

• The soil mixture is then steamed. The steaming time will depend on the quantity of steam produced, which is proportional to the size and capacity of the boiler. A good general criterion is to continue to steam for about 15 minutes after the steam billows the covering tarpaulin. Soil thermometers placed in the corners of the trailer are helpful to judge the correct period for steaming. One minute at 100°C or 10 minutes at 83°C is usually sufficient for controlling soilborne pathogens. Steaming has never been found toxic or harmful to plants grown in UC soil mix with the above composition.

Fertilization

The initial mix contains both macro- and micro-nutrients added during mixing. The micronutrients are tied up in the peat moss. The peat, which acts as a chelating agent, releases sufficient small amounts of micronutrients to the plants for up to one to two years (Nauer, Roistacher and Labanauskas, 1967; 1968).

Liquid fertilizer is applied with each watering using a proportioning device. There are a number of such devices, which inject fertilizers in proportion to the water used. An effective, simple and very inexpensive device is a Venturi-type siphon. Before use the siphon should be calibrated, for many such devices vary considerably from the advertised ratio of concentrate to water as printed in the instructions. To calibrate the siphon, put a measured amount of water (500 or 1 000 ml) in a graduated cylinder, then place the suction end of the siphon into the cylinder and measure the final amount of water exiting from the hose. Allow the water to fill a container until the 500 or 1 000 ml of measured liquid is siphoned up. Then measure the water in the container and convert the results to a ratio.

Another device that injects a given quantity of liquid fertilizer into the water system at a uniform rate in direct proportion to the water flow is the Smith Measuremix proportioner.[1] This is a highly reliable, precision instrument. However, it should be calibrated in the same manner as the Venturi siphon.

A liquid fertilizer formula based on that given by Nauer, Roistacher and Labanauskas (1968) is: 9 parts ammonium nitrate + 3.75 parts calcium nitrate + 2.75 parts potassium nitrate or potassium chloride. This fertilizer should be well mixed and applied at the rate of 67.5 g of mixture to 100 litres of water.

With the UC system of soil mix, the soil should be fertilized directly after mixing and before and immediately after planting since the basic mix contains no nitrogen or potassium. As a general practice, potted plants in a UC system need to be watered periodically with enough volume to flush out any accumulated salts, thereby preventing salinity buildup. It is important that the soil not be filled to the top of the container; a space of 2 to 3 cm should be left between the top of the container and the soil level. This will allow a sufficient volume of water to flush the soil in the container adequately.

---

[1] Information on the Smith Measuremix proportioners can be obtained from Smith Precision Corp., 1299 Lawrence Drive, Newbury Park, California, USA.

REFERENCES

Baker, K.F., ed. 1957. *The UC system for producing healthy container-grown plants.* Calif. Agric. Exp. Sta. Extension Service Manual 23. Repr. 1985. Chipping Norton, Australia, Surrey Beatty and Sons.

Nauer, E.M., Roistacher, C.N. & Labanauskas, C.K. 1967. Effect of mix composition, fertilization and pH on citrus growth in UC-type potting mixtures under greenhouse conditions. *Hilgardia*, 38: 557-567.

Nauer, E.M., Roistacher, C.N. & Labanauskas, C.K. 1968. Growing citrus in modified UC potting mixtures. *Calif. Citrogr.*, 53: 456, 458, 460-461.

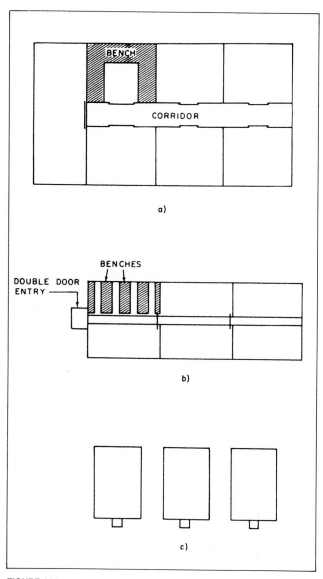

FIGURE 145
Three designs for the layout of plant laboratory greenhouses
a) A six-room design; each room an individual cubicle with a central walkway
b) A three-room design with a central walkway
c) Three separate small greenhouses with individual controls for cool, moderate and warm temperatures

FIGURE 146
An inexpensive wood-and-fibreglass greenhouse at Riverside,
California, with two evaporator coolers

FIGURE 147
The interior of the above greenhouse showing the wood-and-fibreglass
structure. Excellent plants were grown in this inexpensive house using
a UC system of soils, fertilizers and temperature control

FIGURE 148
The Rubidoux laboratory at Riverside, California, has a double-door
entryway and three compartments as in Figure 145b

FIGURE 149
A large fibreglass and aluminium-frame greenhouse. The screened portion at the far end is made of 32-mesh plastic screen to filter the outside air as it passes through cooling cells. Two large fans at the opposite end control air movement

FIGURE 150
The interior of the fibreglass greenhouse at Moncada, Spain, used for large-scale citrus indexing. Note the central walkway, metal benches with plywood top, and cooling cells at the opposite end of the house

FIGURE 151
A small aluminium-and-glass house at Riverside, California, with two
internal heaters and evaporator coolers on the outside

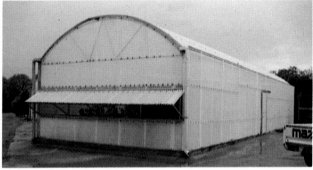

FIGURE 152
A small fibreglass house at Nelspruit, South Africa, used for indexing.
This house is cooled by cooling cells at one end and a fan at the other

FIGURE 153
Greenhouse at Catania, Italy, with top and side vents for
cooling and wire mesh over the structure to prevent
damage to the glass from hail

FIGURE 154
A double-walled polythene greenhouse at Riverside,
California. The double layer gives excellent insulation, and
warm or cool air can be pumped into the space between the
polythene layers for additional heating or cooling

FIGURE 155
Two screenhouses at Bari, Italy, used for growing sanitized grapevine
nuclear stocks. Note the two-door entryway

FIGURE 156
Interior of one of the above screenhouses.
Grapevines are grown in pots that are
individually drip irrigated. Note the gravel floor

FIGURE 157
Wooden benches and concrete block supports at Riverside, California. Note the gravel floor

FIGURE 159
Wooden runners to support large containers, keeping them off the ground. Note the gravel floor

FIGURE 158
Another view of wood and concrete block benches at Riverside, California

FIGURE 160
Benches made of concrete at Campinas, Brazil

FIGURE 161
Steel and wire mesh benches, Mildura, Australia. Note the concrete floor

FIGURE 163
Plastic bench tops on a wooden frame set on concrete blocks in a screenhouse at Lake Alfred, Florida

FIGURE 162
A steel mesh bench on concrete blocks at the USDA greenhouse, Orlando, Florida

FIGURE 164
Spraying wooden benches with copper naphthenate solution as a wood preservative and disinfectant

FIGURE 165
Soft plastic containers suitable for growing rooted grapevine cuttings

FIGURE 166
Various types of plastic and clay pots for growing
grapevines

FIGURE 167
Artificial lighting in a climatized greenhouse at Bari, Italy,
where herbaceous hosts are grown

FIGURE 168
A small heating unit outside the structure at Riverside, California. Heating units placed outside the greenhouse are preferable since they minimize risk of ethylene damage to plants by leakage of exhaust fumes from faulty heaters

FIGURE 169
Two large heaters outside the Rubidoux greenhouse. The warm air is carried by ducts into the upper part of the house and distributed internally by fans and plastic tubing

FIGURE 171
Large perforated plastic tube running the length of the greenhouse. The warm air is blown through the tubing and forced out through the many holes in the tubing

FIGURE 170
A large internal heating unit with a perforated plastic tube attached. Note the circular hole in the plastic tube

FIGURE 172
A separate fan attached near the heater. This fan can be activated independently of the heater fan by a separate thermostat. This permits the circulation of air within the greenhouse for the distribution of latent heat residing in the soil, floors and structure, and is an energy-saving system

FIGURE 173
A large fan at one end of a greenhouse creates negative pressure in the house and forces in the outside air through cooling pads at the opposite end

FIGURE 175
A system used at the Riverside laboratory for filtering incoming air using spun glass and charcoal filters

FIGURE 174
A thermostat control panel with four thermostats for four levels of control. Thermostats independently control fans for the introduction of outside air or for pumping water over the cooling cells. They also control heating by circulating the inside air or by turning on heaters

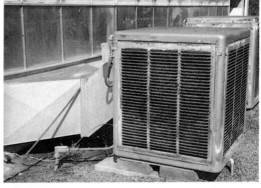

FIGURE 176
A standard commercial evaporator cooler. This cooler is effective if humidity is low. It is relatively inexpensive and, if properly maintained by replacing pads and by periodical cleaning and painting, is effective in cooling greenhouses

FIGURE 177
The same cooler as in Figure 176 but with the side panel removed to show the main squirrel-cage fan, the water pump and water being pumped from the top and down over a pad of wood shavings. The outside air forced through the wet pads is cooled by evaporation

FIGURE 179
A battery of cooling cells at one end of a greenhouse. Cooling by evaporation is most efficient when humidity is low

FIGURE 178
Cooling cells consisting of specially treated cardboard units placed together to make a continuous wall. Water is pumped from a reservoir tank below ground (bottom left) to the top, and drips freely down over the cells by gravity. Cooling is by evaporation

FIGURE 180
Cooling by electric refrigeration units. This is necessary when humidity is too high for evaporation cooling to be effective

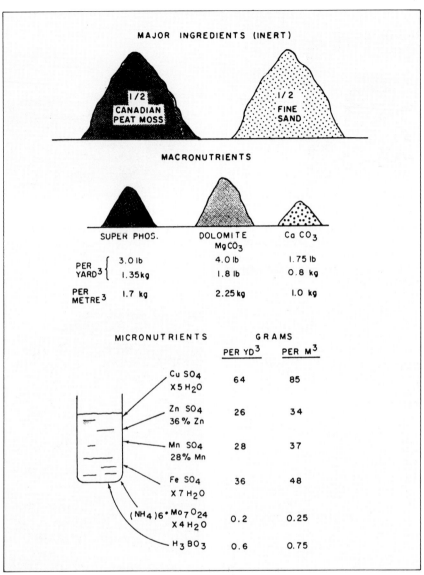

FIGURE 181
The ingredients and fertilizers used in the modified University of California (UC) mix suitable for growing citrus and grapevines

FIGURE 182
A trailer fitted with perforated pipes on the bottom for conducting steam. The top of the trailer is covered with a tarpaulin prior to steaming

FIGURE 183
Flats and pots can be steamed directly in the trailer

FIGURE 184
A fixed steam chamber used to steam containers

Indexing on *Vitis* indicators

G.P. Martelli, V. Savino and B. Walter

The object of indexing is to detect the presence of viruses or virus-like agents in grapevine accessions or selections whose sanitary status is to be assessed.

The use of *Vitis* indicators is compulsory for virus-like and latent diseases, since it represents the only possible way known to date for ascertaining the occurrence of infections. Identification is based on the differential reactions of the indicators.

Indexing programmes, regardless of whether they are carried out on a small or large scale, i.e. for experimental or commercial purposes, require adequate facilities in terms of greenhouses, sheds and land. Availability of nursery land is especially important for growing mother indicator vines (producers of cane wood for indexing) and indicators that have been graft-inoculated. These, irrespective of the inoculation procedure used, except for green grafting, must be grown in the field for no less than two to three seasons to allow a clear expression of symptoms.

INDICATORS

The main indicator plants used for grapevine indexing are listed in Table 6. All indicators are liable to be infected by the whole range of graft-transmissible disease agents, but the symptoms they express may be specific enough for a reliable diagnosis of given diseases, e.g. 110 R for vein necrosis, *Vitis rupestris* for rupestris stem pitting, fleck and asteroid mosaic, red-berried *Vitis vinifera* cultivars for leafroll, LN 33 for corky bark, enations and LN 33 stem grooving and Kober 5BB for Kober stem grooving.

Indicators for European nepoviruses other than GFLV, American nepoviruses and other minor diseases may differ from those in Table 6. These indicators, when known, are mentioned in the descriptions of individual diseases.

For a fairly complete routine indexing, the use of the first seven indicators in Table 6 is advisable. Since at least three individuals of each indicator should be inoculated, each candidate vine under indexing requires a total of 21 grafts.

ESTABLISHMENT AND CARE OF MOTHER VINE INDICATORS

Mother vine indicator plantings constitute the source of cane wood for indexing. Errors in the establishment and care of these plantings may jeopardize indexing programmes.

Mother vine plots should meet as many of the following criteria as possible:
- be located on grounds reasonably close to the research unit in charge of indexing;
- be established on good quality, well-drained and clean soil, preferably with no grapevine history or at least free of grapevines for at least 15 years;
- be separated by at least 20 m from other vineyards to minimize contaminations from adjacent plots by irrigation water, flooding and cultivation;
- be large enough to accommodate other optional indicators in addition to those used routinely.

Before planting, the land must be carefully surveyed for the presence of nematodes, paying attention to virus vector species. Soils infested

TABLE 6

Main indicators for virus and virus-like diseases of the grapevine

Indicator	Disease identified
Vitis rupestris St George	Fanleaf, fleck, asteroid mosaic, rupestris stem pitting
Vitis vinifera Cabernet franc, Pinot noir and other red-berried cultivars	Leafroll
Kober 5BB (*Vitis berlandieri* x *Vitis riparia*)	Kober stem grooving
LN 33 (Couderc 1613 x *Vitis berlandieri*)	Corky bark, enations, LN 33 stem grooving
Baco 22A	Stunting component of leafroll
Vitis riparia Gloire de Montpellier	Vein mosaic
110 R (*Vitis rupestris* x *Vitis berlandieri*)	Vein necrosis
Vitis vinifera Mataro or Mission seedling 1	Leafroll, yellow speckle

with virus vectors, especially *Xiphinema index*, are unsuitable for growing indicator vines even after nematicide treatment. Chemical control with any suitable nematicide product is recommended before planting and before each replanting.

The land should be prepared for good culture and the vine stand arranged for simple maintenance and irrigation, preferably drip irrigation. Although mother vines can be trained according to local systems, bilateral horizontal cordons seem quite suitable (Figures 185 and 186).

Special attention is to be paid to phylloxera, as it not only can endanger self-rooted *V. vinifera* indicators but can cause much damage to leaves and shoots of indicators of American *Vitis* species (Figure 187). Early removal of spring leaves that bear the first galls slows down further reproduction of phylloxera, thus keeping the population low. Spraying schedules for chemical control of phylloxera and possible airborne vectors (pseudococcid mealybugs and leaf-hoppers) should be devised according to necessity and local conditions.

COLLECTION AND STORAGE OF WOOD FOR INDEXING

Mature canes of the current season are collected from indicator mother vines in autumn or early winter. The canes are selected according to size; those with a diameter smaller than 0.8 to 1 cm or larger than 2 to 2.5 cm are discarded. The selected canes are cut into 40- to 50-cm long pieces bearing four to six nodes. Cuttings are bundled in groups of 20 to 30 and labelled (plastic tags marked with a pencil or felt pen are quite suitable). Individual bundles are wrapped in cloth or moist paper and then in a polythene bag, or are placed in slightly moist peat moss and wrapped in a polythene sheet (Figure 188). The bundles are labelled and stored at 2 to 4°C. Although not strictly necessary, dipping cuttings before packing in 1 percent aqueous suspension of Captan or other suitable fungicide may prove useful to prevent moulding. Cold-stored cuttings keep for up to three years and may be withdrawn at any time for use.

Cane wood of donor vines to be indexed is collected, treated and stored in the same way. Well preserved cold-stored canes can be used throughout the year for ELISA testing for nepoviruses and closteroviruses.

PREPARATION OF INDICATOR WOOD FOR INDEXING

Canes of the indicators are withdrawn from cold storage and washed clean prior to being cut into smaller pieces of a size suitable to the type of graft inoculation used. Indicator wood can be used within a few hours of withdrawal from cold storage. No special precautions are required for its handling.

GRAFTING TOOLS AND MATERIALS

Grafting tools may be those used locally, provided that they are of good quality and are kept clean and sharp. The major tools and materials are (Figure 189):

- pruning shears;
- budding knife;
- budding rubber tape;
- plastic wrapping tape or plastic film;
- raffia (for field grafting).

GRAFT INOCULATION METHODS

Several grafting methods can be used for indexing, their choice being dictated by available facilities and expertise and by the extent of the indexing programme. Each procedure has its own advantages and drawbacks but all, once mastered, guarantee high graft takes, a step of utmost importance for successful indexing.

Field (cleft) grafting

Rooted or unrooted cuttings of the candidate vine are planted in the nursery at close spacing (about 30 cm apart) in groups of three in a row and are allowed to grow undisturbed for a whole season. The following winter (Figures 190 and 191) the young vines are pruned (Figures 192 and 193), the soil around them is removed (Figures 194 and 195) and the roots from the crown are cut away (Figures 196 and 197). The vines are then cut back so as to leave a stump with a smooth surface (Figures 198 and 199). The canes of the indicator to be used as scions are cut into one-bud pieces 10 to 15 cm long, which are placed in boxes and kept moist (Figure 200). Bud sticks of the scion with an extremity cut in a wedge shape (Figure 201) are inserted in a 2- to 3-cm-deep cleft made in the stumps with a budding knife (Figures 202 to 204). The graft is then tied with raffia (Figures 205 and 206) and covered with moist, well pulverized soil (Figures 207 and 208). The graft will callous through the winter but will remain dormant until the next spring (Figure 209). With many diseases, symptoms will appear in the first flush of vegetation (Figure 210), four to six weeks after bud burst.

A distinct advantage of field grafting is that there is no transplanting crisis, which minimizes losses of grafted indicators. However, the procedure is slow and quite expensive and requires fairly large land surfaces and skilled operators.

Chip-bud grafting

Canes of indicator vines are withdrawn from cold storage just before use and are cut into cuttings of two to three nodes. The cuttings are planted with the basal part about 6 cm deep in moist sand, directly on a bench heated at 25 to 27°C or in trays placed on the heated bench (Figure 211). When rooted (Figure 212), the cuttings are potted in clay or plastic containers and kept in a greenhouse at 22 to 24°C for a couple of weeks before grafting. During this period rooted cuttings do not require special care if the environment is clean and the potting mixture was properly sterilized before use.

Wood of candidate vines to be indexed is removed from storage the day before its use and held at room temperature (20 to 22°C). For grafting, a bud (or a chip of wood from the internode) is removed from the cane of the donor vine by making a cut above the bud, then a cut at an angle just below the bud, letting the knife travel smoothly upward through the wood until it reaches the first cut. A notch of a size compatible with that of the chip taken from the candidate vine is made on the indicator just below the top shoot. The chip is placed in the notch, tied with budding rubber and wrapped with plastic tape or plastic film (Figure 213).

Grafted cuttings can be held in a greenhouse for symptom development or placed in a shed for a hardening-off period of two to three weeks. Before they are transferred to the field the wrapping tape is removed and the inoculum checked. Unsuccessful grafts are replaced by rebudding below the first graft. When transplanting out of doors, ensure that the whole set of indicators for each candidate is planted along the same row, with a spacing of about 30 cm between rootings.

Chip-bud grafting is a very handy method; it is simple and easy to perform, does not require special skill and ordinarily gives excellent results, provided that the indicators are properly handled, especially during transplanting and hardening-off.

Machine (bench) grafting

Several types of mechanical grafting equipment are available, from the very simple hand- and foot-operated (Figures 214 to 218) to the more sophisticated and costly heavy-duty and electricity-powered machines (Figures 219 to 221).

Examples of machine-made grafts are illustrated in Figures 222 and 223. In the former a chip with a bud excised from a donor vine is fitted into a compatible notch carved in the indicator cutting and tied; in the latter a one-bud, 5- to 6-cm-long scion (usually the indicator) and a 30- to 40-cm-long disbudded cane of the candidate vine used as understock are joined by an omega, V-shaped or saw-type cut. Immediately after grafting the graft union can be waxed for protection, although this is not strictly necessary. This is done by quick-dipping the grafted cane tip into a melted grafting wax (Figures 224 and 225) or, more simply, into melted, low melting point (54 to 56°C) paraffin.

Grafted dormant cuttings are stratified in a box with moist sawdust or peat moss (Figures 226 to 228) and placed in a hot room at about 30°C for callousing. When callousing is completed all around the graft union (usually within four to five weeks) (Figure 229), the boxes are placed in a cooler, well-lit room or in an open shed for three to four additional weeks for hardening-off of grafted rootings (Figure 230) before transplanting in the field.

Machine grafting is quick and is thus especially suitable for large-scale indexing. However, several steps of the procedure, such as callousing and transplanting, are critical and may cause substantial losses of grafted rootings. Additional losses because of low graft take are to be expected because of the nature of the material processed, which may be weakened by infection or may contain pathogens that interfere with graft union. Even under the best conditions not more than 70 percent of grafts are successful, and this percentage may drop dramatically with less compatible scion/stock combinations.

Green grafting

In principle, green grafting is a simple technique, suitable for both small- and large-scale greenhouse indexing.

For small-scale indexing the stocks, which can be either indicators or candidate vines, are forced to grow vigorously to allow development of one robust shoot (Figure 231). This shoot is cut at the level of the third or fourth internode

(Figure 232). Green scion material the same size as that of the stock is collected (Figure 233) and grafted by making a slanting cut that fits into a comparable slant on the stock or by inserting a wedge-cut shoot tip into a cleft made in the shoot of the stock (Figure 234). The graft is tied with plastic grafting tape or plastic film (Figure 235) and protected from dehydration with a polythene bag (Figure 236). The bag is removed after a couple of weeks and the graft is left on the greenhouse bench for appearance of symptoms. For large-scale indexing, a mechanized green grafting technique has recently been developed in France by INRA (Institut national de la recherche agronomique, Colmar) and GCEV (Groupement champenois d'exploitation viticole, Mumm Recherche, Epernay) (Walter *et al.*, 1990).

Herbaceous cuttings (18 to 35 cm long and 1.5 to 2.5 mm in diameter) are collected from grape indicator varieties and grown either in a mixture of soil compost, sand and peat (1:1:1) or on rockwool cubes (Figure 237) watered twice a week with the following nutrient solution (Huglin and Julliard, 1964):

KNO_3, 800 g

$MgSO_4 \cdot 7H_2O$, 300 g

$(NH_4)_2 \cdot HPO_4$, 200 g

H_2SO_4, 50 ml

H_3PO_4, 17 ml

$MnSO_4 \cdot H_2O$, 1.5 g

H_3BO_3, 1.5 g

$ZnSO_4 \cdot 7H_2O$, 1.5 g

$CuSO_4$, 0.5 g

$(NH_4)_6 \cdot Mo_7O_{24} \cdot 4H_2O$, 0.05 g

Edetate sodium Fe (e.g. Sequestrene 138 FeR), 15 g

The chemicals are dissolved in 1 000 litres of water and pH is adjusted to 6.0.

Cuttings are grown at a mean temperature above 20°C and a day length of 16 hours with a light intensity of 2 500 lux.

Wedge-shaped, one-bud cuttings are inserted into a V-shaped cleft made at the top of the cuttings of the indicator by means of an appropriate grafting machine (Figure 238) (Martin *et al.*, 1987). Scions and stocks are held together by small plastic clothes-pegs (Figure 239). Each grafted green cutting is then inserted with the basal extremity into a rockwool cube (Figure 240) and maintained in a plastic forcing chamber under saturated humidity at 25°C for at least 18 days. The cubes are watered twice a week with the above nutrient solution. The grafts are moved from the forcing chamber to greenhouse boxes at temperatures most appropriate for the expression of symptoms of the various diseases (e.g. 22°C for leafroll and fleck, 26°C for vein necrosis). Symptoms of fanleaf, leafroll (Figure 241), fleck, corky bark (Figure 242), stem pitting (Figure 243), vein mosaic (Figure 244), vein necrosis (Figure 245) and graft incompatibility appear much more quickly (20 to 60 days, except for stem pitting, which requires three to five months) than with conventional indexing techniques.

GROWTH CARE OF GRAFTED INDICATORS

Regardless of whether grown in greenhouse, screenhouse or open field, indicators should be forced to grow vigorously and be protected from diseases and pests which may obscure symptoms and even endanger their survival. For field-grown indicators, cultural practices are the same as those routinely used in nurseries.

Symptoms are usually read twice a year: once in late spring or early summer for growth abnormalities (leaf and cane deformations, reduced vigour, stunting), necrotic or chromatic disorders of the foliage (vein necrosis, chlorotic mottling, yellow discolorations), wood pitting or grooving; and once in autumn for abnormal pigmentation (red or yellow) of the leaves and, after leaf shedding, for cane abnormalities. At the end of the indexing period (usually three years) the vines are uprooted, the scion is cut down to 20 to 25 cm from the bud union, the roots are cut away and the main axis containing part of the scion and part of the rootstock is decorticated for observing the presence of stem pitting or stem grooving symptoms. To facilitate peeling, the material is autoclaved at 120°C for 20 to 30 minutes.

Records of indicator reactions should be kept for each candidate accession or selection subjected to indexing. The time and appearance of symptoms, their type and severity are noted and compared with the responses of positive controls, i.e. reference indicator vines inoculated with budwood from sources known to be affected by specific diseases and grown at random in the indexing plot. Each plot should also contain negative controls, i.e. healthy, non-inoculated indicator vines serving as trap plants for diseases that may spread naturally in the indexing nursery.

REFERENCES

Huglin, P. & Julliard, B. 1964. Sur l'obtention de semis de vignes très vigoureux à mise à fruits rapide et ses répercussions sur l'amélioration génétique de la vigne. *Ann. Amélior. Plantes*, 14: 229-244.

Martin, C., Vernoy, R., Carré, M., Vesselle, G., Collas, A. & Bougerey, C. 1987. Vignes et technique de culture *in vitro*. Quelques résultats d'une collaboration entre recherche publique et entreprise privée. *Bull. OIV*, 675/676: 447-458.

Walter, B., Bass, P., Legin, R., Martin, C., Vernoy, R., Collas, A. & Vesselle, G. 1990. The use of a green grafting technique for the detection of virus-like diseases of the grapevine. *J. Phytopathol.*, 128: 137-145.

FIGURE 185
Newly established mother vine planting of
indicators. Note the drip irrigation pipe

FIGURE 188
Collection and storage of wood: a bundle of mature
four- to six-node-long canes collected in early winter;
the bundle placed in a polythene sheet and covered
with moist peat moss; the wrapped bundle ready for
cold storage

FIGURE 186
Mother vine planting of indicators
in full growth

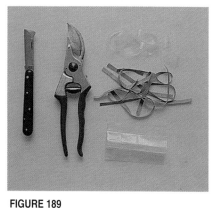

FIGURE 189
Tools and materials for grafting:
budding knife, pruning scissors,
plastic budding tape, budding rubber
ties, plastic film

FIGURE 187
Phylloxera galls on the
underside of leaves of an
American rootstock. Heavy
attacks of this aphid may
seriously affect the growth of
indicator vines and reduce the
yield of budding wood

FIGURE 190
General view in winter of a planting of candidate vines
to be indexed

FIGURE 193
Pruning is completed

FIGURE 191
Close-up of a group of three candidate vines to be
indexed. Note the close spacing between vines

FIGURE 194
Soil around candidate vines is removed to expose
crown roots

FIGURE 192
Candidate vines undergo pruning for eliminating
the shoots

FIGURE 195
Crown roots are exposed

FIGURE 196
Cutting away crown roots

FIGURE 199
Stumps of candidate vines ready for cleft cutting

FIGURE 197
Candidate vines ready for beheading

FIGURE 200
Wooden box containing bud
sticks of an indicator and a
bundle of raffia

FIGURE 198
Candidate vines being beheaded

FIGURE 201
Indicator bud stick wedge-cut at one extremity

FIGURE 202
Median cleft being made with a budding
knife in the stump of a candidate vine

FIGURE 203
Indicator bud stick being inserted by its
wedge-shaped extremity into the cleft of
the candidate vine

FIGURE 204
Indicator bud stick in place

FIGURE 205
Graft being tied with raffia

FIGURE 207
Grafted vines being covered with soil

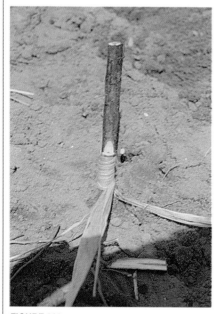

FIGURE 206
Tying of the graft completed

FIGURE 208
Soil covering completed. End of grafting operation

FIGURE 209
Spring growth pushing from grafts

FIGURE 212
Vegetating canes ready for chip-budding and transplanting

FIGURE 210
Indexing plot in summer

FIGURE 213
Chip-bud grafting: indicator rooting with a notch below
the top shoot and a chip from the candidate vine
next to it; the chip is inserted into the notch, tied with
plastic budding tape and wrapped with plastic film

FIGURE 211
Canes of indicator vines forced to root on a heated bench.
The canes are stuck into trays containing sterilized
river sand

FIGURE 214
Simple hand-operated grafting
machine (V-shaped cut)

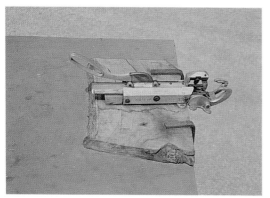

FIGURE 215
Hand-operated bench grafting machine (chip-bud graft)

FIGURE 217
Foot-operated bench grafting machine
(omega cut)

FIGURE 216
Hand-operated bench grafting machine (V-shaped cut)

FIGURE 218
Close-up of the machine illustrated in Figure 217

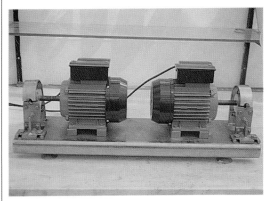

FIGURE 219
Saw-type electrlcity-powered grafting machine

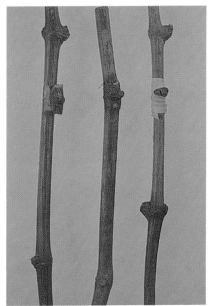

FIGURE 222
Outcome of machine chip-bud grafting:
non-rooted cane of a candidate vine with a
machine-made notch and a bud from the
indicator vine being inserted; bud in place
after insertion in the notch; graft tied with
budding rubber band

FIGURE 220
Omega-type electricity-powered grafting machine

FIGURE 221
Omega-type machine in operation
(Photo: B. Di Terlizzi)

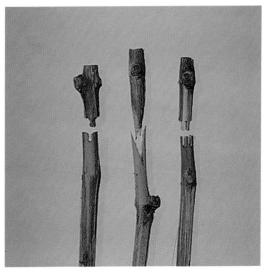

FIGURE 223
Types of machine-made cuts (from left to right):
omega, V-shaped, saw-type

FIGURE 224
Waxing newly made machine grafts
(Photo: B. Di Terlizzi)

FIGURE 225
Commercial grafting wax before melting

FIGURES 226, 227 and 228
Successive steps of the stratification in sawdust of
grafted cuttings for callousing
(Photo: B. Di Terlizzi)

FIGURE 229
Calloused graft ready for hardening
(Photo: B. Di Terlizzi)

FIGURE 230
Green hardened grafts ready for transplanting in the field
(Photo: B. Di Terlizzi)

FIGURE 231
Green grafting: candidate vine pushing
a shoot

FIGURE 232
Green shoot cut at the level of the third
internode

FIGURE 233
Green cuttings and a shoot tip from
an indicator ready for green grafting.
Note the wedge-shaped extremity

FIGURE 234
Green grafting performed. The indicator shoot
has been inserted into a cleft cut made in the
shoot of the candidate vine

FIGURE 235
Graft union wrapped with plastic film

FIGURE 236
Green-grafted vine protected with a polythene bag
to ensure high humidity conditions until the graft
has taken

FIGURE 237
Large-scale green grafting. Young self-rooted
indicator vines growing on rockwool cubes

FIGURE 238
Machine for green grafting

FIGURE 239
Green graft held in place by a plastic clothes-peg

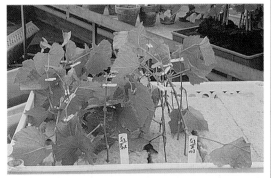

FIGURE 240
Green-grafted cuttings stuck in rockwool, ready for
forcing

FIGURE 241
Leafroll symptoms shown by a
green-grafted indicator (left)

FIGURE 242
Incipient corky bark symptoms shown by a green-grafted indicator

FIGURE 243
Stem pitting symptoms shown by a green-grafted indicator

FIGURE 244
Vein mosaic symptoms shown by a green-grafted indicator

FIGURE 245
Stunting induced by vein necrosis in a green-grafted indicator

Use of herbaceous hosts

G.P. Martelli

Many of the viruses infecting grapevines in nature (e.g. those listed in Table 2) are transmissible by inoculation of sap to herbaceous hosts. Various closteroviruses differ in that some are mechanically transmissible, although with varying degrees of difficulty, whereas others are not. All phloem-limited viruses with isometric particles have so far resisted sap transmission. The procedure for mechanical inoculation is simple. It can be applied with a reasonable degree of success if some basic requirements in the management of herbaceous hosts and handling of the inoculum are fulfilled.

COLLECTION AND HANDLING OF INOCULUM

In principle, parts of any living organ of an infected vine can serve as a source of inoculum for sap transmission. In practice, however, the chances for successful transmission are best if young, tender tissues from developing leaves or root tips are used. When leaves are used, as most frequently happens, the terminal parts of symptomatic shoots are collected (Figures 246 and 247) and placed in a polythene bag (Figure 248), which is sealed and labelled (Figure 249). If the place of collection is at a distance from the laboratory, the samples should be kept in an ice chest (Figures 250 and 251) for transport. Exposure to direct sun or to a hot environment such as a car boot should be avoided. In the laboratory the samples are processed immediately or placed in a refrigerator at 4°C. Cold storage can last up to a couple of weeks or more if the samples have been properly handled

during collection and transport. Prolonged storage in a freezer at -20°C is possible but not devoid of risk, for the particles of certain viruses (e.g. nepoviruses) may disassemble during thawing, thus decreasing the infective power of the extracts.

TOOLS AND MATERIALS FOR INOCULATION

Tools and materials for inoculation (Figure 252) include the following:
- sterile porcelain pestles and mortars (sterilized by heating in a stove, autoclaving or prolonged boiling);
- abrasive powder (celite or 500-mesh carborundum);
- cotton swabs or any other suitable implement for rubbing and spreading the inoculum on the leaves;
- tap water;
- extraction medium.

For grapevine viruses two main extraction media are used. The first is nicotine (2.5 ml) dissolved in distilled water (97.5 ml). This solution can be kept for several months, especially if stored in a refrigerator. The alternative is phosphate buffer, 0.1 M, at pH 7, made from the following stock solution (1 M):
- 136.09 g of KH_2PO_4 in distilled water to 1 litre (solution A)
- 268.077 g of Na_2HPO_4 in distilled water to 1 litre (solution B)

Mix 3.86 ml of solution A with 6.14 ml of solution B and dilute tenfold. This solution does not need to be prepared freshly each time;

it can be stored in the refrigerator for a few weeks.

INOCULATION PROCEDURE

The hosts to be inoculated are selected (Figure 253), transferred to the greenhouse bench where the inoculation is to be made and dusted with the abrasive powder (Figure 254). The infected sample is placed in the mortar and ground with one to three volumes of the extraction medium (Figures 255 and 256). The slurry is gently rubbed (with a cotton swab dipped in the medium) on to the host's leaves (Figures 257 and 258), which are then rinsed with tap water (Figure 259). Inoculated plants are grown at temperatures preferably not below 18°C or above 26°C and are checked for symptom appearance.

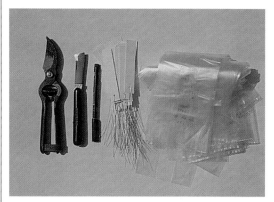

FIGURE 246
Tools and materials for collecting samples (from left to right): pruning scissors, budding knife, felt pen, plastic labels, polythene bags

FIGURE 248
Collected shoot tip being placed in a polythene bag

FIGURE 247
Shoot tip from a symptomatic vine being cut away for collection

FIGURE 249
The bag containing samples is labelled and sealed

FIGURE 250
An ice chest for transporting fresh samples to be tested
in the laboratory

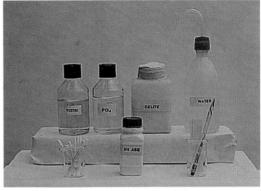

FIGURE 252
Tools and materials for inoculation. Back row: nicotine
(2.5 percent in water), phosphate buffer, celite (abrasive
powder), tap water for rinsing inoculated plants. Front
row: cotton swabs, sodium ascorbate (antioxidant),
pencil and labels

FIGURE 251
Samples placed in the ice chest, ready for
transport

FIGURE 253
Some of the herbaceous hosts most commonly used for
isolating and culturing plant viruses. Upper row, left
to right: *Nicotiana glutinosa, Nicotiana benthamiana,
Nicotiana clevelandii.* Lower row, left to right:
*Chenopodium amaranticolor, Gomphrena globosa,
Chenopodium quinoa*

FIGURE 255
Leaf tissues from a vine to be tested in a sterile mortar
prior to grinding

FIGURE 254
Herbaceous hosts (*Nicotiana tabacum*) ready for
mechanical inoculation. The plant on the left has been
dusted with celite

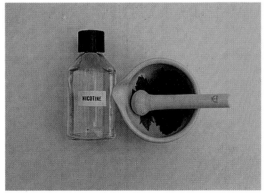

FIGURE 256
Leaf tissues ground in the presence of a suitable
extraction medium

FIGURE 257
Mortar containing inoculum ready to be rubbed on
the leaves of the adjacent host

FIGURE 258
Inoculation by gentle rubbing of the inoculum on the leaf
surface with a cotton swab

FIGURE 259
Host leaves rinsed with tap water immediately after
inoculation

Part III

Laboratory methods for detection and identification of infectious agents

SEROLOGY
Immunoprecipitation
G.P. Martelli

Precipitation or precipitin reactions derive their name from the visible precipitate formed when adequate quantities of antigen and antibodies are allowed to interact.

When two reactants (antigen and antibody) capable of recognizing each other come into contact, a complex develops in the form of a lattice of antigen and antibody molecules. This results from the establishment of bonds between epitopes (i.e. antigenic determinants at the surface of the antigen molecule) and the corresponding antigen-combining sites located at the extremity of the arms of the antibody molecule (IgG, IgM). The antigen-antibody complex becomes insoluble and precipitates.

Precipitin reactions can take place in a liquid medium (liquid precipitin) when the reactants are mixed together or in an agarized medium (immunodiffusion in gel) when originally separated reactants are allowed to diffuse into one another through the pores of the agar.

Liquid precipitin is used for viruses whose particles are too long (500 nm and above) to diffuse in agar. Gel diffusion is used for viruses with particles up to 300 nm long (i.e. Tobraviruses Tobamoviruses, Furoviruses). Viruses with longer particles can be made to react in gel diffusion, provided that their particles are previously dissociated by SDS (sodium dodecyl sulphate) or another strong denaturing agent.

MICROPRECIPITIN TEST

The microprecipitin test is the same as tube precipitin but is more economical, as it requires less antiserum. It is also more sensitive, as small precipitates not detected by the naked eye become visible under a dissecting microscope.

Materials
- glass or plastic Petri dishes 100 mm in diameter. If glass plates are used, their bottoms must be given a coat of silicone paste to be made hydrophobic
- micropipettes or Pasteur pipettes
- paraffin oil
- reactants: antiserum and virus-containing plant extract (usually clarified plant sap)

Procedure
- Put a series of drops of the infected extract in the bottom of the Petri dish.
- Add to each drop an equal amount of antiserum at increasing dilution.
- Mix drops thoroughly.
- Cover with paraffin oil.
- Incubate for 1 hour at 37°C.
- Read reaction under a dissecting microscope. Drops in which a precipitate has formed are scored positive.

IMMUNODIFFUSION (GEL DOUBLE DIFFUSION TEST)

The advantage of the gel double diffusion test is that antigens and antibodies, while diffusing into one another, form a gradient that allows the precipitin reaction to occur where the ratio between the reactants reaches the right proportion. Furthermore, since different antigens diffuse at different rates and precipitate at different sites, gel diffusion tests reveal the presence of mixed antigens and allow their separation and the assessment of relationships between them.

Materials (Figure 260)
- glass or plastic Petri dishes 55 or 100 mm in diameter. Glass plates need to be made hydrophobic with a coat of silicone paste
- Pasteur pipettes
- gel cutters (cork borers) or preformed templates for well patterns
- gel medium composed of 7 to 15 g bacteriological agar or agarose, to be dissolved by heating in 1 litre of 0.2 M buffer (phosphate, citrate) or saline (0.85 percent sodium chloride), with the addition of a preservative (e.g. 0.02 percent sodium azide)
- reactants: antiserum and infected crude or clarified plant sap

Procedure (Figure 261)
- Place Petri dishes on a flat surface. Pour warm medium to form a layer about 3 mm thick and leave to solidify at room temperature. Plates can be stored in a plastic bag in a refrigerator (4°C) until needed.
- Choose the desired well pattern. A very common pattern consists of a central well 2 to 3 mm in diameter for the antiserum, surrounded by eight peripheral wells 4 mm in diameter for the antigens, each 4 to 5 mm from the edge of the central well.
- Cut cylinders of agar with the gel cutter, following the chosen pattern, and remove agar plugs with a needle or by suction.
- Place reactants in the wells.
- Cover Petri dish and leave to stand at room temperature or at 4°C until precipitin lines are formed. Although the reaction may begin to appear as soon as 3 hours after filling the wells with reactants, precipitin lines are fully developed after 12 to 14 hours.

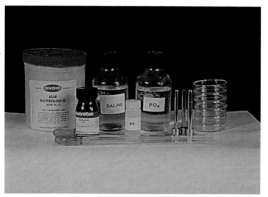

FIGURE 260
Materials and tools needed for gel double diffusion tests.
Background: bacteriological agar, buffered solution
(PO$_4$) or physiological solution (saline), glass or plastic
(55 or 100 mm diameter) Petri dishes. Foreground:
Pasteur pipettes, a preservative (sodium azide),
antiserum and cork borers to be used as gel cutters

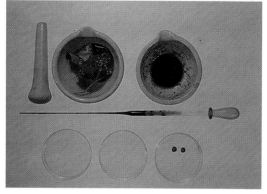

FIGURE 261
Procedure for gel diffusion tests (from left to right and top
to bottom): leaf samples to be tested serologically in a
porcelain mortar prior to grinding; leaf tissues ground
preferably without addition of extraction media; a Pasteur
pipette containing crude sap to be loaded in the wells made
with a cork borer in the agarized medium contained in the
plastic Petri dishes

SEROLOGY
Enzyme-linked immunosorbent assay (ELISA)
S.M. Garnsey[1] and M. Cambra

This guide to the use of enzyme-linked immunosorbent assay (ELISA) is based on the chapter on ELISA in the handbook *Detection and diagnosis of graft-transmissible diseases of citrus,* but with a few minor changes for specificity to grapevine. It describes several common variations of ELISA. Some background information is presented to help the user understand the technique and make modifications to this highly flexible procedure for specific applications. Information on the selection of techniques, on preparation of samples for testing and on the basic steps of the ELISA protocol is provided. The materials, reagents and equipment needed are indicated, and some specific examples are given.

Properly used, ELISA is a sensitive, accurate and rapid detection method. It is especially effective when large numbers of samples must be assayed, when results are needed rapidly and when suitable indicator plants and/or greenhouse facilities are not available. ELISA has been developed for grapevine fanleaf virus and a number of other mechanically and non-mechanically transmissible viruses such as grapevine virus A, leafroll-associated closterovirus I, II and III and fleck-associated isometric virus, as well as for flavescence dorée MLOs and *Xylella fastidiosa,* the agent of Pierce's disease.

ELISA is simple and can be carried out by most people after brief training and some practice. As with any indexing procedure, some experience is necessary to use ELISA accurately and confidently. New users should consult several of the excellent general references on ELISA (e.g. Clark and Bar-Joseph, 1984; Clark, Lister and Bar-Joseph, 1988; Sanchez-Vizcaino and Cambra Alvarez, 1987) which provide additional details on theory and application. It is very useful to visit a laboratory where ELISA is practised in order to observe the procedure and to practise it under the guidance of an experienced user. Begin with a well-known system, and study the effects of adjusting reactant concentrations and test conditions.

Extensive training and background in serology and immunology are not essential to use ELISA, but understanding of some basic concepts is necessary. ELISA is a serological technique, and like other serological procedures it is based on the concept that many proteins are antigenic when injected into animals and that the immunized animal will form antibodies to them.

These antibodies can be obtained from the serum of the immunized animal and will react specifically with the antigen to which they were formed. A primary requirement to begin ELISA is a useful source of antibody to the pathogen to be detected. This in turn means that antigen specific to the pathogen must be identified and

---

[1] Mention of a trademark, warranty, proprietary product or vendor does not constitute a guarantee by the United States Department of Agriculture and does not imply its approval to the exclusion of other products or vendors that may also be suitable. This disclaimer is a prerequisite for the contribution of United States Government employees.

purified sufficiently to produce the needed antibodies. Antigen purification and antibody production are beyond the scope of this section, but information on these topics is contained in some of the references cited (Clark, Lister and Bar-Joseph, 1988; Van Regenmortel, 1982).

Also fundamental to ELISA is the concept that various enzymes can be bound to antibody molecules to form a conjugated molecule that has enzymatic activity and is also serologically active. Since enzymes are highly active and can be detected at low concentrations, they are effective labels. Enzyme-labelled antibodies can be detected when they are exposed to a substrate that enzymes can change. Normally, a substrate that changes colour as a result of the enzyme action is used. The amount or rate of colour change can then be used to measure the amount of antibody present. Enzyme labels provide a sensitivity similar to that of radioactive labels and have several important advantages: they are stable, inexpensive and safe to use, and they can be used successfully without sophisticated equipment.

The enzyme label may be attached directly to the antibody used to detect the antigen in question (the detecting antibody). This is called a direct assay, of which the highly popular double antibody sandwich technique described below and illustrated in Figure 262a is a good example. The label may also be used indirectly. In this case, the label is attached not to the detecting antibody, but rather to a second antibody specific to the detecting antibody. Antibodies of one species are antigenic when injected into an animal of a second species. For example, rabbit immunoglobulins can be injected into another animal such as a goat to create a goat anti-rabbit antiserum. These goat anti-rabbit antibodies are useful to detect antibodies from rabbits that were originally prepared to detect another antigen.

Indirect assays are more sensitive and also avoid the need to prepare a conjugate to each antibody used. Several forms of indirect assay are described in the following section and are also illustrated in Figure 262 (b and d). The relative advantages of direct and indirect systems are discussed in the following section.

Several other molecular interactions are frequently used in conjunction with ELISA, either to purify immunoglobulins or to amplify reactions and increase sensitivity. Protein A is a cell wall component of the bacterium *Staphylococcus aureus* and has the unique characteristic of binding to the immunoglobulin protein of many mammalian species. The binding site is on the Fc region of the immunoglobulin and not on the antigen binding site. Protein A is frequently used to purify antibodies by affinity chromatography. It can also be conjugated with enzymes and used in assays to detect immunoglobulins.

A second important system is the biotin/avidin system. Biotin, a small vitamin, has a very high affinity for avidin, a 68 000 molecular weight glycoprotein. Antibodies and enzymes can be conjugated with several molecules of biotin to form a "biotinylated" molecule. Each avidin molecule has four binding sites for biotin. This multiplying interaction has been exploited in several ways to amplify the number of enzyme molecules associated with each antigen-bound antibody and thereby increase sensitivity. One example is illustrated in Figure 262c.

Another fundamental concept for ELISA is that proteins such as antibodies and virus coat proteins will adsorb strongly to the surface of certain plastics such as polystyrene and polyvinyl chlorides. Protein binding also occurs to some forms of cellulose nitrate. These materials are frequently referred to as "immunosorbents" or the "solid phase" in ELISA protocols. The protein binding to immunosorbent materials is not specific and is not a serological reaction such as occurs between antigen and antibody molecules. If a mixture of antibodies is exposed to an immunosorbent plastic, all will bind. Similarly, when a crude extract from a diseased plant is placed in an ELISA plate, proteins of the pathogen and proteins of the host present in the extract will both be bound.

Binding either the antibody or the antigen component of a serological system to a solid phase is very useful because the bound component can subsequently be used to probe complex mixtures of potential reactants. Only those that are serologically related will be trapped. All non-reactive components can then be removed by washing and will not interfere with subsequent steps. For example, when an extract from a virus-infected plant is placed in the wells of a microtitre plate coated with antibodies to that virus, virus antigens in the extract will be bound to the trapping antibody and all non-related proteins will be removed by the subsequent washing step.

Undesired adsorption of antibody or antigen proteins to the plastic can be avoided by using non-ionic detergents such as Tween 20 in incubating solutions or by adding an excess of a non-specific protein to block all sites not occupied by the desired serological component. For example, the buffer used to coat plates with trapping antibody does not contain Tween 20, but Tween 20 is incorporated in subsequent steps where any non-specific binding of other proteins should be avoided.

Immunoblotting procedures are not specifically discussed in this section. However, much of the information and the general con-cepts presented are directly applicable to immunoblotting procedures. The main differences are that the solid phase for immunoblotting is usually cellulose nitrate, the substrate used to measure presence of the antigen-antibody-enzyme complex is different and incubation conditions may be somewhat modified.

ELISA PROCEDURES
Numerous variations of the ELISA procedure can be devised (Clark, Lister and Bar-Joseph, 1988; Engvall and Pesce, 1978; Jones and Torrance, 1986; Koenig and Paul, 1982; Maggio, 1980). The selection depends on the sensitivity, specificity and convenience required; the presence of interfering factors; and the type and activities of the antisera available. The basic steps for four commonly used variations of ELISA are outlined here and illustrated in Figure 262. In three of the variations, a, b and c, the solid-phase (ELISA plate) is coated with antibody to the antigen to be detected. This antibody, identified as trapping antibody (TA in the figure), then traps its corresponding antigen (identified as V) from suspension or solution. In the fourth variation (Figure 262d), the antigen (V) is trapped directly on the solid phase and detected with its specific antibody.

Double antibody sandwich
The inexperienced user should start with the double antibody sandwich (DAS) where possible. This has been the most commonly used form of ELISA for plant virus detection since its description by Clark and Adams (1977). The components of DAS are illustrated in Figure 262a. The immunosorbent surface is a plastic microtitre plate with wells designed for ELISA as shown in Figure 263. A dilute solution of unlabelled antibody is added to the wells of the plate (Figure 264), and the antibody adsorbed on the plastic becomes the trapping antibody (TA) as illustrated in Figure 262a. After washing to remove any excess antibody (see Figure 277), the sample (antigen) is added as shown in Figure 265. Antigens (V in Figure 262) specific to the

bound trapping antibody attach themselves to it, but other proteins remain in solution and are removed by washing. The antigen attached to the trapping antibody is detected by adding a labelled antibody (LA in Figure 262a) specific to the antigen (Figure 266). The label is the enzyme (E) previously conjugated to the antibody. When substrate specific to the enzyme is added in the final step (Figure 267), a colour develops as a result of enzyme action (Figures 268 to 270). The amount of colour and its rate of development are correlated to the amount of labelled antibody bound to the antigen which had been trapped by the antibody attached to the plate.

DAS can be done with a single good quality polyclonal antiserum. The immunoglobulins present are partially purified, and one portion is saved for use as trapping antibody while another is conjugated to an enzyme. Alkaline phosphatase is commonly used as the enzyme and the conjugation can be done in the presence of dilute glutaraldehyde (Clark, Lister and Bar-Joseph, 1988). The antibodies for coating and detection do not have to come from the same source, e.g. monoclonal antibodies could be used for coating and a polyclonal antiserum could be used to prepare the enzyme-labelled antibody.

Double antibody sandwich indirect

DAS can be converted to an indirect procedure (DAS-I), which is illustrated in Figure 262b. The first two steps are the same as in DAS. However, the antigen bound to the trapping antibody is detected by an unlabelled intermediate antibody (IA) which is specific to the same antigen but is from an animal species different from the one used to prepare the trapping antibody. For example, if the trapping antibody was prepared in rabbits, the detecting or intermediate antibody could be from a mouse or a chicken. The unlabelled IA which attaches to the antigen is detected by an enzyme-labelled antibody (LA)

specific to the IA. Because the IA is from a different species than the TA, the LA binds only to the IA and no non-specific binding of the LA to the TA occurs. The amount of LA is measured by adding substrate and measuring colour change as in DAS.

DAS-I ELISA involves an additional step (Figure 262b) but is more sensitive and also allows use of a commercially prepared enzyme-labelled antibody to the IA. A single LA can also be used for multiple virus detection systems. In addition, the intermediate antibody does not have to be purified and is needed in only a limited quantity. If the intermediate antibody is highly specific, e.g. most monoclonals, then a highly specific antiserum is not required for coating. The major problem is that antibodies to the same antigen must be prepared in two different animals. If the trapping and the intermediate antibodies are from the same species, the labelled antibody used to detect the intermediate antibody will also bind to the trapping antibody and result in a non-specific response.

A system has been devised to carry out DAS-I using a single antiserum (Adams and Barbara, 1982; Clark, Lister and Bar-Joseph, 1988). To do this, the antibodies are treated with the enzyme pepsin to remove the Fc portion of the molecule. The remaining F(ab')$_2$ fragment still has the antigen binding sites and will bind to the immunosorbent, but will not bind to protein A. The F(ab')$_2$ fragments are used as trapping "antibody" and the whole antibody is used as the intermediate antibody. Enzyme-conjugated protein A is then used instead of a labelled antibody to detect the intermediate antibody. It does not react to the trapping "antibody" because the Fc region has been removed.

The DAS-I procedure can be further modified to amplify the reaction achieved. This is commonly done using a biotin-avidin interaction in which the labelled antibody is biotinylated to react with avidin molecules conjugated to multiple enzyme molecules, as illustrated in Figure 262c. Different types of amplification are possible and special kits may be purchased to perform them. Users should be aware of the possibility of employing amplification when additional sensitivity is needed, but regular procedures should be tested before amplification is attempted.

Plate-trapped antigen

Another basic approach to ELISA is the plate-trapped antigen procedure (Figure 262d). The approach is to trap the antigen (V) on the plastic surface, then react the trapped antigen with an unlabelled intermediate antibody (IA) specific to it. The IA is then detected as in DAS-I using an enzyme-labelled antibody (LA) specific to the IA. This procedure, called plate-trapped antigen indirect (PTA-I) ELISA, is relatively simple and involves no advance purification of antisera or conjugate preparation if a commercially prepared enzyme-labelled anti-body to the unlabelled IA is used. The PTA-I procedure is usually less sensitive than DAS or DAS-I for use with crude plant extracts and may not be effective when antigen concentration in the sample is low. Since binding to the plate is non-specific, the target antigen and other proteins present in the extract compete for the available binding sites on the plate. Plate-trapped antigen tests can be conducted as a direct assay using an enzyme-labelled antibody to the antigen, but sensitivity is even lower than for the indirect method, and the conjugate must still be prepared. Amplification procedures as described for DAS-I can also be used for the PTA-I procedure to increase sensitivity.

The specific steps and schedules for these types of ELISA are described in Schedules 1 to 3.

SAMPLING

Selection of appropriate samples for testing is critical. Although ELISA is a sensitive procedure, reliable results may not be obtained if poor samples are tested. Virus titre in grapevine tissue often varies markedly, and thousandfold differences in antigen concentration can occur over a relatively short period. Virus concentrations are generally highest in young, expanding flush tissues. They decrease rapidly as tissues mature under hot-weather conditions and more slowly under cool conditions. Avoid sampling old, mature tissue during the summer months in hot climates unless preliminary testing indicates that reliable samples can be taken. If the virus or pathogen is phloem-limited (e.g. the grapevine pathogens leafroll, rugose wood-associated closteroviruses and fleck-associated isometric virus), then the tissue sample collected must contain phloem tissue. Older bark tissue can be sampled if the cambium is active, but generally it is less reliable than young shoot flush, bark or young leaf midribs (Figures 271 and 272). Young root tips may be useful under some conditions.

A composite sample from several sites on the vine should be collected; normally three to five locations per vine are sampled. Increase sampling if the pathogen is irregularly distributed or when trying to monitor a recent infection.

Fresh tissue can normally be stored for at least seven to ten days at 4°C when kept in a plastic bag or sealed container.

Samples can also be stored frozen at -20°C or below for extended periods. Unprocessed fresh tissue should be used; or tissue can be diced, placed in extraction buffer and frozen in the grinding tube (Figure 273). The sample should not be ground prior to freezing because fresh extracts often lose much activity when frozen. Frozen samples should not be stored in an automatically defrosting freezer. Extracts can be stored for long periods when freeze-dried; this is a good way to store a source of consistent reference (control) samples. Always test a storage method with the specific pathogen under study in order to prove its effectiveness.

EXTRACTION

Numerous buffers and different additives have been used for extraction of tissue samples with different virus-host systems (Bar-Joseph and Garnsey, 1981; Clark, 1981; Clark and Bar-Joseph, 1984; Clark, Lister and Bar-Joseph, 1988; McLaughlin *et al.*, 1981). Phosphate-buffered saline (PBS) or 0.05 M Tris, pH 7.5 to 8.0, without any additives usually give good results for sandwich assays. Additives such as polyvinylpyrrolidone (PVP), EDTA and DIECA are generally unnecessary for citrus, but with grapevine tissues 2.5 percent nicotine or 2 percent PVP may be useful (see chapter on grapevine degeneration – fanleaf). Test the effect of additives before using them routinely. For plate-trapped antigen procedures, try extraction of the sample in carbonate coating buffer, pH 9.6, or in 0.05 M Tris, pH 8.0. Do *not* use Tween 20 in the extraction buffer for samples to be plate-trapped.

Normally, the ratio of buffer to sample tissue should be at least 1:10. Higher concentrations of tissue may actually reduce reaction rates and make sample preparation more difficult.

There are many ways to grind samples. Pestle and mortar are fine for small numbers of tender samples. Addition of an abrasive, such as fine sand or carborundum, to the sample or powdering the tissue in liquid nitrogen makes grinding easier. A dispersion homogenizer (Figure 274) equipped with a 10 to 25 mm diameter shaft is a good choice when large numbers of samples are to be processed. A 2 to 10 ml sample can be rapidly ground in a test-tube or centrifuge tube of suitable diameter and length with this type of homogenizer. Fibrous tissue, such as bark and

leaf midribs, should be cut into short pieces (2 to 5 mm) prior to grinding or the shaft will become clogged with fibre and need to be cleaned between samples. Two rinses of the grinder shaft in 500 to 1 000 ml clean water are usually adequate (Figure 275). Run the homogenizer briefly in each rinse solution.

Chill samples prior to grinding to offset heating during the grinding process. It is normally not necessary to keep the sample on ice during grinding, unless unusually long grinding is required and the sample becomes warm to the touch. Frequent users of dispersion homogenizers should wear earplugs to protect their hearing, and homogenizers should be isolated.

If necessary, samples can be prepared with very minimal equipment. When virus concentration is high, extensive disruption of the sample tissue is usually unnecessary. One method is simply to place a small piece of tender tissue directly in buffer in the well of an ELISA plate with forceps and then squeeze it to release the cell contents. Tissue can also be crushed in a small plastic bag using a mallet or smooth stone and the extract moved by pipette into the test plate.

Samples containing a lot of debris after extraction can be difficult to pipette. Remedies include centrifugation of the sample to pellet the debris or filtering the sample through a coarse filter such as cheesecloth or glass wool (Figure 276). Cutting off a portion of the tapered tip of a plastic pipette creates a wider orifice and is often quick and effective. It is frequently quicker to rinse a repeating pipette between samples than to change tips, so only a limited number of tips need be modified.

WASHING

Proper washing of the plates between steps is important. The standard procedure is to wash the plate three times between each step with phosphate-buffered saline (PBS) containing 0.5 percent Tween 20 (PBST). Sodium azide is frequently included in PBST solutions as a preservative. However, it is highly poisonous and may form an explosive complex with some metals. Sodium azide is unnecessary in ELISA wash solutions and should be omitted. The two most critical wash steps are after sample incubation when cross-contamination must be avoided between wells containing different samples (omit the first wash immediately if carry-over between wells occurs) and after the conjugate incubation step. If even a minor residue of unattached conjugate remains, high background readings may occur. (Add another wash at this point when in doubt.) Various plate washers are available which can promote consistent washing operations, but a plastic squeeze bottle will work well for small volumes of plates (Figure 277). Solutions in the plate wells can be removed by aspiration to avoid contamination, but usually the plate is inverted rapidly with a quick shake of the hand and tapped firmly on clean blotting paper or paper towels.

TEST CONDITIONS

A wide variety of reactant concentrations and incubation times and conditions have been reported for ELISA (Clark and Bar-Joseph, 1984; Clark, Lister and Bar-Joseph, 1988; McLaughlin *et al.*, 1981). The choice of conditions depends to some extent on the basic goals. With high concentrations of reactants, short incubation times can be used, and if necessary the entire ELISA procedure can be completed within two hours. Increasing the incubation time while decreasing concentration (especially of the conjugate) will conserve reactants. Reactions occur most rapidly at 30 to 37°C, but room temperature (20 to 28°C) will give satisfactory results. Gentle shaking during incubation may improve efficiency. Many workers find it convenient to do the sample incubation step overnight and often do this at 4 to 6°C.

Some experimentation will be necessary to determine optimum conditions for each situation. Schedules 1 to 3 give examples which should provide a good starting point. Moderate changes in times and conditions are unlikely to cause a test failure, and changes can often be made to render the schedule more convenient for the user with no loss of information. New users should certainly experiment with different schedules to find the optimum for their purpose.

One of the major variables in ELISA to be evaluated is the concentration of conjugate to use. Commercially prepared enzyme-labelled antibodies normally have a recommended working dilution (frequently between 1/1 000 and 1/2 000). Conjugates that are prepared experimentally may differ markedly, and published values for other systems are of little help. Optimum dilutions of 1/100 to 1/20 000 of the stock preparation (approximately 1 mg per ml) have been reported. The effective dilution will depend on the basic affinity of the antibody, the titre of specific and non-specific (host) antibodies, the source and activity of the enzyme used and the effectiveness of the conjugation procedure. When starting with a new or unknown batch of conjugate, test three tenfold dilutions starting at a 1/100 dilution to determine the approximate activity. Using these results, make a second test in the appropriate dilution range indicated. Normally, the objective is to obtain a strong positive reaction to a good positive sample within 20 to 60 minutes and little or no reaction to healthy extracts. If the conjugate concentration is so high that a reaction is instantly visible, a background reaction is often also observed with healthy extracts (and even with buffer controls). Reduce conjugate concentration and, if a non-specific reaction persists, adsorb the antiserum against a concentrated extract of healthy plant tissue to remove antibodies against healthy antigens. If possible, do this before purifying the IgG. Adding healthy tissue extract to the buffer used to dilute the conjugate may also reduce non-specific reactions (Clark, Lister and Bar-Joseph, 1988).

The use of appropriate controls is essential. Each plate should have at least one healthy and one known positive sample as controls. A buffer control is also useful to determine the level of background reaction to healthy extracts. Frequently, a slightly higher reading will be observed for the buffer control than for the healthy extract because proteins in the extract block exposed protein binding sites on the plastic, which may later non-specifically bind conjugate molecules. Each sample should be tested in at least two wells. A random loading pattern can be used, but paired wells are normally used for routine work. Special applications may require additional replication and randomization.

Edge effects in the plates were frequently noted when ELISA first became popular and outer wells were avoided. Plates have steadily improved and normally all wells can be used. Uniformity in new lots of plates can be checked by loading a uniform sample in all wells.

RECORDS

One of the major tasks of any indexing procedure is to identify samples properly during the testing process and to record results in a usable format.

The identity of the sample must be maintained through the multiple steps of collection, processing, extraction and testing. It is usually convenient to give each sample a code number at the time of collection and to use this code during the test process. If samples are collected directly in the grinding vessel (usually a glass or plastic tube), labelling steps can be reduced. If the sample is collected in a container other than the grinding vessel, a transferable label is often convenient. Many ELISA plates have a coding system on the plate margins to identify individual wells, but there is no space to mark individual wells on the plate. Most workers develop a data sheet similar to the one shown as Figure 279, which is used to record the loading sequence of test samples and other pertinent data such as reactant concentration and incubation conditions. Each plate and data sheet should have a corresponding number recorded in a logbook to facilitate retrieval of information.

It is important to mark the loading pattern for each plate prior to loading samples and to arrange the samples to be loaded in the appropriate sequence. Note changes or errors that may occur during loading, and store tubes and samples under refrigeration until the testing process is complete.

Visual readings of the plate can be recorded directly on the plate data sheet. Printouts from a plate reader (Figure 270) can also be attached to the original data form. Use of computers to store and analyse data is increasing rapidly. Computers are convenient for long-term storage of large amounts of data. Data from the reader may also be converted into another format for further analysis and spreadsheet presentation on the computer. Permanent visual records of important tests can also be obtained by photographing individual plates on a light box or over a white background. A 35-mm transparency film is economical, and a standard exposure can be obtained if a constant light source is used.

EVALUATION OF RESULTS

Evaluation of ELISA results often presents some problems to the novice user. With highly specific antisera and samples with good antigen titre, results are normally very clear. When the antisera used are weak or contain some antibodies to host proteins and/or the samples have a very low antigen titre, determination of a positive result can be more difficult. Reactions can be evaluated visually with some precision if background readings for healthy controls are low. Normally, the eye can discern differences in OD_{405} of 0.05 to 0.1 above a low background. A graded scale with three to five levels is useful to report the relative degree of reaction. Where greater accuracy is required, the degree of reaction can be measured by testing a diluted sample in a spectrophotometer or by reading the plate in an ELISA plate reader (Figure 270). A wide variety of plate readers are available, from simple manual models suitable for modest numbers of plates to highly automated models capable of various levels of data analysis and storage.

Recently, emphasis has increased on measuring reaction rate rather than a single final optical density value. This eliminates some sources of error where accurate quantitative data are needed and also allows more accurate comparison of samples with large differences in antigen concentration. Rate calculation requires several measurements of the same plate at a measured time interval, and a plate reader is essential. Some plate readers do rate calculations automatically. Expensive plate readers are not necessary until a definite need for them is identified.

The limits of reliable detection are correlated to the precision used and the number of replications. Confidence levels can be calculated statistically when in doubt. Most workers establish an arbitrary threshold value relative to the healthy control for a positive reaction, such as a reading twice that of the healthy control or the healthy reading plus 0.1 OD. Where reaction to healthy extracts is low (<0.05 OD), the eye can usually consistently detect reactions of 0.1 OD or higher and this becomes the effective limit for visual recording. It is best to establish a conservative threshold rating and retest all questionable samples. Some experimentation with different dilutions of known samples will help.

PURIFICATION OF IMMUNOGLOBULINS

Numerous procedures are now available for purification of immunoglobulins (IgG) from polyclonal antisera or for purification of monoclonal antibodies from culture fluid or ascites fluid. The easiest method is to use a commercial kit or system containing detailed instructions. Many of the kits are based on separation of the IgG component by protein A bound to a solid substrate. The IgG is subsequently eluted from the protein A by changing buffers and collected. An alternative, which is slower but inexpensive, is to precipitate the IgG from solution by use of ammonium sulphate and then fractionate the dialysed resuspended pellet by column chromatography on a DEAE cellulose column (Figure 278).

Purified IgG solutions are normally adjusted to a concentration of 1 mg per ml (equal to an OD_{280} of 1.4 when measured spectrophotometrically) and stored at 4°C in PBS containing at least 0.02 percent sodium azide. IgG preparations can also be stored in 50 percent glycerol at -20°C or freeze-dried. More detailed instructions can be found in Clark, Lister and Bar-Joseph (1988).

PREPARATION OF ENZYME-LABELLED ANTIBODIES

Conjugated molecules of antibody and enzyme can be prepared in several ways (Bar-Joseph and Garnsey, 1981; Clark and Adams, 1977; Clark and Bar-Joseph, 1984; Clark, Lister and Bar-Joseph, 1988; Engvall and Pesce, 1978). Alkaline phosphatase is the most widely used enzyme, and the single-step glutaraldehyde method is commonly used to prepare alkaline phosphatase conjugates. Test this system first before experimenting with other procedures. Alkaline phosphatase is usually obtained commercially as an enzyme precipitate in a salt solution. The precipitate is recovered from solution by centrifugation and about 5 mg is dissolved in 2 ml of a 1 mg per ml solution of IgG. The mixture is dialysed thoroughly to remove excess salts and then fresh glutaraldehyde is added to a final concentration of 0.05 percent. After 4-hour incubation, the conjugate is dialysed three times to remove the glutaraldehyde and stored at 4°C in PBS containing 0.04 percent sodium azide and 5 mg per ml bovine serum albumen. Use an enzyme source with a quality suitable for ELISA. Careful dialysis is very important for good results. New lots of conjugate must be tested to determine optimum dilutions for use (Sanchez-Vizcaino and Cambra Alvarez, 1987). Good-quality conjugates can normally be used at at least 1/500 dilution and dilutions as great as 1/5 000 or more may be possible. See Clark, Lister and Bar-Joseph (1988) for further details and for preparation of horseradish peroxidase conjugates.

Conjugates are stable for long periods at 4°C. Freezing or freeze-drying are not recommended unless preliminary testing indicates it is possible. If freezing is necessary, add 50 percent glycerol.

BASIC SUPPLIES AND EQUIPMENT

Sources of supplies and equipment change rapidly and new equipment continues to appear. It is not possible to list all suppliers in every country. To be of help, we have indicated some possible sources for some essential supplies. It is not essential to use these specific sources, and the user may, in fact, find a more convenient and economical local source than those listed. The *Laboratory buyer's guide*, available from International Scientific Communications, Inc., PO Box 870, Shelton, CT 06484, USA, is a useful directory of manufacturers. Basic equipment and supplies are emphasized rather

than some of the more sophisticated and expensive equipment also available for large-scale clinical work. It is assumed that most users of the latter already know the sources of supply.

A source of antiserum to the pathogen you wish to detect is essential. Both polyclonal and monoclonal antisera have been produced to a number of grapevine pathogens, and more are being developed. Polyclonal antisera are frequently preferable for general detection work where discrimination of a particular isolate is either unnecessary or undesirable. Unless the antigen used to produce the antiserum was well purified, polyclonal antisera frequently contain some antibodies to plant proteins as well as to the specific pathogen. It is advisable to check the specificity of the antiserum to be used in the initial stages to ensure acceptability. Monoclonal antibodies are generally more specific because, if properly prepared, only a single epitope of the antigen is involved. To do DAS-I assays, antibodies are needed from two animal species unless the $F(ab')_2$ procedure is followed.

Production of high-quality antisera is frequently a time-consuming and difficult task, especially for people without experience in this process. Inexperienced users of ELISA should ordinarily try to obtain a small amount of antisera from existing sources for their initial work. Frequently, a modest supply of antiserum can be obtained from a colleague working on grapevine pathogens. Most grapevine virologists are members of the International Council for the Study of Viruses and Virus Diseases of the Grapevine (ICVG) and can be identified by writing to the Secretary, c/o Federal Agricultural Research Station of Changins, 1260 Nyon, Switzerland. Small amounts of purified IgG or enzyme-labelled conjugate may also be available for limited experimental tests.

Increasing access to antisera from commercial sources and the American Type Culture Collection can be expected. Current commercial outlets of ELISA supplies for grapevine pathogens include: Agdia Inc., 30380 County Road 6, Elkhart, IN 46514, USA; Ingenasa, Hermanos Garcia, Noblejas 41, 28037 Madrid, Spain; Sanofi Santé Animale, Z.I. de la Ballastière, BP 126, 33500 Libourne, France; Bioreba AG, Gompenstrasse 8, 4008 Basel, Switzerland; Boehringer, Box 310120, 6800 Mannheim, Germany; and Loewe Biochemia GmbH, Residenza Betulle 801, 20090 Segrate (MI), Italy.

If a prepared ELISA kit or purified sources of IgG and labelled conjugates are available, ELISA tests can be done in a very simple laboratory equipped with a balance, a simple pH meter, basic glassware, a refrigerator and a supply of deionized water. The other essential items are ELISA plates, several repeating pipettes and the chemicals to prepare the necessary buffers (see below). To purify IgG and to prepare antibody-enzyme conjugates, it is necessary to have access to a low-speed centrifuge, a UV spectro-photometer and some column chromatography equipment.

ELISA plates are available from numerous suppliers including Dynatech Laboratories, 14340 Sullyfield Circle, Chantilly, VA 22021, USA and Nunc Inc., 2000 N. Aurora Road, Naperville, IL 60540, USA. Plate quality can vary. Plates that work well for sandwich assays may work less well for plate-trapped antigen assays. If possible, find a reputable local supplier and test several types of plates before buying a large supply.

Several repeating pipettes, as shown in Figures 264 and 265, are more or less essential. Fixed volume models are economical and will suffice, but adjustable models are much more convenient and can be used for many other tasks. The minimum requirement is one pipette that will measure accurately in the 1 to 20 µl range (to make dilutions of IgG and conjugates) and one

that will operate in the 100 to 1 000 µl range (or 200 to 1 000 µl). Pipettes for the ranges of 20 to 200 µl and 1 to 5 ml are also extremely useful. Multichannel pipettes (Figures 264 and 266) that can simultaneously dispense the same volume into 4 to 12 wells are very useful when many plates must be loaded. All repeating pipettes use disposable plastic tips. If possible, select pipettes that can use interchangeable tips. There are numerous manufacturers of pipettes and numerous models. Consult a laboratory supply company or manufacturers such as Flow Laboratories S.A., Lugano, Switzerland or Rainin Instruments Co., Woburn, MA 01801, USA for current information.

Plates can be read visually, but if large numbers of plates are to be done on a regular basis a plate reader greatly speeds reading and makes evaluation of results easier. Numerous models with varying degrees of automation are available and details change rapidly. It is not necessary, and probably not advisable, to buy a reader until the user has some initial experience and knows the system will work. Before purchase, ask for a demonstration and also consult users in other laboratories for their recommendations. Large-scale users should consider readers that are computer compatible.

See Part IV for additional information on laboratory equipment.

ELISA BUFFERS AND SOLUTIONS

A limited number of chemicals are required to make the buffers and solutions needed for ELISA, and these are shown in Table 7. Many of these should be readily available in most biological laboratories. Specific sources and catalogue numbers have not been listed, but suggestions can be obtained from other ELISA users. Sigma Chemical Co., St Louis, MO 63178, USA; Boehringer Mannheim Biochemicals, Indianapolis, IN 46250, USA; and Pharmacia

TABLE 7
List of chemicals for ELISA

1.	Alkaline phosphatase type VII	
2.	Bovine serum albumen	BSA
3.	DEAE cellulose	
4.	Diethanolamine	$NH(CH_2CH_2OH)_2$
5.	Glutaraldehyde	$OCH(CH_2)_3CHO$
6.	Hydrochloric acid	HCl
7.	Ovalbumen	
8.	4-Nitrophenyl phosphate	
9.	Polyvinyl pyrrolidone MW 40 000	
10.	Potassium chloride	KCl
11.	Potassium phosphate	KH_2PO_4
12.	Sodium azide	NaN_3
13.	Sodium bicarbonate	$NaHCO_3$
14.	Sodium carbonate	Na_2CO_3
15.	Sodium chloride	NaCl
16.	Sodium hydroxide	NaOH
17.	Sodium phosphate (dibasic)	Na_2HPO_4
18.	Tris(hydroxymethyl)aminomethane HCl	Tris-HCl
19.	Tween 20	

LKB Biotechnology AB, Uppsala, Sweden or Piscataway, NJ 08854, USA are useful general sources for chemicals, enzymes and antibodies mentioned in this section if satisfactory local suppliers cannot be located. See also the *Laboratory buyer's guide* mentioned above.

The materials listed are for the alkaline phosphatase system. If horseradish peroxidase or another enzyme is used, make appropriate changes in substrate and substrate buffer.

The formulae for buffers and substrate solutions needed are shown in Table 8. Use glass-distilled or high-quality deionized water to prepare buffers and solutions. The formulae given in Table 8 are for one-litre quantities. Larger quantities of the wash buffer (PBST) than of other solutions are used. The dry salts needed for PBS can be weighed in advance in units to make a convenient volume, mixed dry and stored in sealed plastic

bags until needed. A new supply of PBS can be obtained rapidly, as needed, by adding the required volume of distilled water to the weighed salts.

Use standard buffer solutions with pH values near 7.0 and 10.0 to calibrate pH meters. Store buffers (except PBST) at 4°C if possible.

SCHEDULES

Schedules are shown here for types of ELISA illustrated in Figure 262, a, b and d, to provide some specific examples of reactant concentration and incubation time and conditions. The details shown are those typically used with an alkaline phosphatase enzyme-label system. No schedule is shown for the amplified form of DAS-I illustrated in Figure 262c. The preliminary steps are the same as used for DAS-I and the schedule for the enhancement steps will vary with the

TABLE 8
ELISA buffers and solutions

1. **Coating buffer**

   1.59 g Na_2CO_3

   2.93 g $NaHCO_3$

   0.2 g NaN_3

   (pH should be 9.6)[1]

2. **Phosphate buffered saline (PBS)**

   8 g NaCl

   0.2 g KH_2PO_4

   2.9 g $Na_2HPO_4 \cdot 12\ H_2O$

   (1.15 g anhydrous)

   0.2 g KCl

   (pH should be 7.2 to 7.4)[1]

3. **Washing buffer (PBST)**

   1.0 litre PBS

   0.5 ml Tween 20

4. **Extraction buffer**

   1.0 litre PBST

   20 g polyvinyl pyrrolidone, MW 40 000[2]

   (Option: 15.7 g Tris-HCl, adjusted to pH 7.8 with NaOH)

5. **Conjugate buffer**

   1.0 litre PBST

   20 g polyvinyl pyrrolidone, MW 40 000[2]

   2.0 g ovalbumen

   0.2 g NaN_3

6. **Substrate buffer**

   97 ml diethanolamine

   0.2 g NaN_3

   (adjust pH to 9.8 by adding HCl)

7. **Reaction stopping solution**

   120 g NaOH

---

[1] pH should be close to value, adjust slightly if necessary.
[2] Polyvinyl pyrrolidone is not essential for extraction or conjugate buffers.

enhancement protocol used. Specific instructions are generally provided with the enhancement materials when these are purchased in kit form.

Normally, 200 µl of solution are added per well, but smaller volumes can be used. If NaOH is used to stop the reaction, 50 µl are added to the wells already containing substrate.

Schedule 1

Double antibody sandwich ELISA

1. Coat ELISA plates (Figure 264) with antibodies (IgG) diluted to 1 to 2 µg per ml in carbonate coating buffer. Incubate for 1 to 4 hours at 25 to 30°C and wash three times with PBST (Figure 277).

2. Add sample extracts (Figure 265) prepared at a 1/10 to 1/20 dilution in extraction buffer in duplicate or triplicate wells. Incubate for 2 to 4 hours at 30 to 37°C or overnight at 4 to 6°C. Wash thoroughly three times with PBST. Avoid cross-contamination of samples when washing.

3. Add enzyme-antibody conjugate diluted in conjugate buffer to an optimum concentration (normally between 1/500 and 1/5 000) (Figure 266). Incubate 2 to 4 hours at 37°C. Wash at least three times with PBST to remove unbound conjugate from the wells.

4. Add substrate freshly prepared at a concentration of 0.6 to 1 mg per ml in substrate buffer (10 percent diethanolamine, pH 9.8) (Figure 267). Incubate until strong colour change develops in positive controls (normally 30 to 60 minutes) (Figure 268) and read plates (Figure 270). Plates may be read at several intervals without stopping the reaction, to calculate reaction rate, or the reaction can be stopped at an appropriate time by addition of 3 M NaOH and a single reading can be made. If plates are read visually, score the estimated relative strength of reaction. If read on a spectrophotometer or with a plate reader (Figure 270), record the OD_{405} values.

Schedule 2
Double antibody sandwich indirect ELISA

1. Coat ELISA plates (Figure 264) with antibodies (IgG) specific to the antigen to be tested. The IgG concentration should be 1 to 2 µg per ml in carbonate coating buffer. Incubate for 1 to 4 hours at 25 to 30°C and wash three times with PBST (Figure 277).

2. Add sample extracts (Figure 265) prepared at a 1/10 to 1/20 dilution in extraction buffer. Load duplicate or triplicate wells with each sample. Incubate for 2 to 4 hours at 30 to 37°C or overnight at 4 to 6°C. Wash thoroughly three times with PBST, avoiding cross-contamination of samples.

3. Add unlabelled intermediate antibody at an appropriate dilution, normally 0.25 µg per ml or less. Incubate for 30 to 60 minutes at 30 to 37°C, and wash plate three times with PBST.

4. Add enzyme-labelled antibody specific to the intermediate antibody, diluted according to the instructions supplied. Incubate 1 to 2 hours at 30 to 37°C and wash plate carefully at least three times with PBST.

5. Add substrate freshly prepared at 0.6 to 1 mg per ml concentration in substrate buffer (10 percent diethanolamine, pH 9.8). Incubate until strong colour change develops in positive controls (normally 30 to 60 minutes) and read plates. Plates may be read at several intervals without stopping the reaction so rate of reaction can be calculated, or the reaction can be stopped at an appropriate time by addition of 3 M NaOH and a single reading can be made. If plates are read visually, score estimated relative strength of reaction. If read on a spectrophotometer or plate reader, record the OD_{405} values.

Schedule 3
Plate-trapped indirect ELISA

1. Add antigen extracts to uncoated ELISA plates and incubate 1 to 4 hours at 25 to 30°C or overnight at 4 to 6°C. (Note that samples prepared for PTA *should not* have Tween 20 in the extraction buffer.) Wash plates three times with PBST and avoid contamination while washing.

2. Add unlabelled antibody specific to the antigen at an appropriate dilution (normally 1 µg per ml or less). Unpurified polyclonal antisera or ascites fluid can be used. Incubate 1 to 2 hours at 30 to 37°C and wash three times with PBST.

3. Add enzyme-labelled antibody specific to the unlabelled antibody, at the specified dilution (normally about 1/1 000), and incubate 1 to 2 hours at 30 to 37°C. Wash carefully at least three times with PBST.

4. Add substrate prepared at a concentration of 0.6 to 1 mg per ml in substrate buffer (10 percent diethanolamine, pH 9.8). Incubate until a strong colour change develops in positive controls (normally 30 to 60 minutes) and read plates. Plates may be read at several intervals without stopping the reaction so rate of reaction can be calculated, or the reaction can be stopped at an appropriate time by addition of 3 M NaOH and a single reading can be made. If plates are read visually, score estimated relative strength of reaction. If read on a spectrophotometer or plate reader, record the OD_{405} values.

TROUBLE-SHOOTING

Several types of problems may be encountered with ELISA. Some understanding of the operating principles of ELISA helps for systematic trouble-shooting to identify and correct the problem. Several of the most common situations are covered here. If the suggestions given do not solve the problem encountered, seek the help of someone who has extensive experience with ELISA.

No reaction or reaction is very slow
The common causes are:
- use of an incorrect buffer in one or more steps;
- inactivation of antigen during processing or storage;
- loss of enzyme activity in the conjugate (commonly occurs if conjugate is accidentally frozen);
- low affinity of antibody for test antigens or loss of affinity when antibody was conjugated;
- inactive substrate;
- gross miscalculation when making dilutions.

Test conjugate and substrate by mixing a small amount of dilute conjugate with fresh substrate in a small beaker. If no reaction occurs, test each separately again and replace the faulty component. Test reactivity of the antibody by an alternative procedure such as immunodiffusion or microprecipitation. Check calculation of dilutions and test a freshly prepared positive control. Test other extraction buffers. Run a different virus system with the same buffers and protocols.

Colour development is non-specific
When all wells, including buffer and healthy controls, show a strong reaction, it could indicate:
- the antibody source for either coating or conjugate phase antigens is giving non-specific reaction (common with antisera to SDS-degraded antigens);
- the washing was incomplete, especially after the conjugate step;
- the coating of the plate was incomplete, or Tween 20 was left out of the PBST buffer and enzyme-labelled antibody is being adsorbed non-specifically to the plate;
- the substrate is contaminated or faulty.

If the buffer control is negative, but the healthy control shows a positive reaction, the antiserum used probably has a high concentration of antibodies to healthy plant antigens. The alternatives are either to absorb the antiserum with healthy plant proteins to remove the antibodies in the serum specific to the healthy plant proteins, or to prepare other antisera.

Colour development is erratic
The common causes of erratic reaction within a plate are:
- defective plates;
- careless performance of one or more steps, especially washing;
- failure to mix thoroughly diluted IgG and conjugate solutions;
- contamination between wells.

Check another source of plates and review the care used in the operating procedure.

Reaction very rapid; some reaction also in healthy samples

This normally indicates that the conjugate concentration is much too high. Try several tenfold dilutions. If differentiation still fails to occur between healthy and positive samples with normal incubation periods, see recommendations above.

REFERENCES

Hundreds of references are available on the ELISA technique and its application to numerous plant viruses. We list a few here. Many of these citations contain additional literature citations.

Adams, A.N. & Barbara, D.J. 1982. The use of F(ab')$_2$-based ELISA to detect serological relationships among carlaviruses. *Ann. Appl. Biol.*, 101: 495-500.

Bar-Joseph, M. & Garnsey, S.M. 1981. Enzyme-linked immunosorbent assay (ELISA): principles and application for diagnosis of plant viruses. *In* K. Maramorosch & K.Γ. Harris, eds, *Plant diseases and vectors: ecology and epidemiology*, p. 35-59. New York, Academic Press.

Bar-Joseph, M., Garnsey, S.M., Gonsalves, D., Moscovitz, M., Purcifull, D.E., Clark, M.F. & Loebenstein, G. 1979. The use of enzyme-linked immunosorbent assay for detection of citrus tristeza virus. *Phytopathology*, 69: 190-194.

Clark, M.F. 1981. Immunosorbent assays in plant pathology. *Annu. Rev. Phytopathol.*, 19: 83-106.

Clark, M.F. & Adams, A.N. 1977. Characteristics of the micro-plate method of enzyme-linked immunosorbent assay for the detection of plant viruses. *J. Gen. Virol.*, 34: 475-483.

Clark, M.F. & Bar-Joseph, M. 1984. Enzyme immunosorbent assays in plant virology. *Methods Virol.*, 7: 51-85.

Clark, M.F., Lister, R.M. & Bar-Joseph, M. 1988. ELISA techniques. *In* A. Weisbach & H. Weisbach, eds, *Methods for plant molecular biology*, p. 507-530. New York, Academic Press.

Engelbrecht, D.J. 1980. Indexing grapevines for grapevine fanleaf virus by enzyme-linked immunosorbent assay. *Proc. 7th Meet. ICVG*, Niagara Falls, NY, USA, 1980, p. 277-282.

Engvall, E. & Pesce, A.J. 1978. *Quantitative enzyme immunoassay*. London, Blackwell Scientific Publications. 129 pp.

Gonsalves, D. 1979. Detection of tomato ringspot virus in grapevines: a comparison of *Chenopodium quinoa* and enzyme-linked immunosorbent assay (ELISA). *Plant Dis. Rep.*, 63: 962-965.

Hampton, R., Ball, E. & De Boer, S., eds. 1990. *Serological methods for detection and identification of viral and bacterial plant pathogens. A laboratory manual.* St Paul, MN, USA, Am. Phytopathol. Soc. Press. 389 pp.

Huss, B., Muller, S., Sommermeyer, G., Walter, B. & Van Regenmortel, M.H.V. 1987. Grapevine fanleaf virus detection in various grapevine organs using polyclonal and monoclonal antibodies. *Vitis, 25: 178-188.*

Jones, R.A.C. & Torrance, L. 1986. *Developments and applications in virus testing*. Wellesbourne, War., UK, AAB. 300 pp.

Koenig, R. & Paul, H.L. 1982. Variants of ELISA in plant virus diagnosis. *J. Virol. Methods*, 5: 113-125.

Kölber, M., Beczner, L., Pacsa, S. & Lehoczky, J. 1985. Detection of grapevine chrome mosaic virus in field-grown vines by ELISA. *Phytopathol. Mediterr.*, 24: 135-140.

Maggio, E.T. 1980. *Enzyme-immunoassay*. Boca Raton, Florida, USA, CRC Press. 295 pp.

McLaughlin, M.R., Barnett, O.W., Burrows, P.M. & Baum, R.H. 1981. Improved ELISA conditions for detection of plant viruses. *J. Virol. Methods*, 3: 13-25.

Permar, T.A., Garnsey, S.M., Gumpf, D.J. & Lee, R.F. 1988. A monoclonal antibody which discriminates strains of citrus tristeza virus. *Phytopathology*, 78: 1559.

Sanchez-Vizcaino, J.M. & Cambra Alvarez, M. 1987. *Enzyme immunoassay techniques, ELISA, in animal and plant diseases*. Tech. Ser. No. 7, 2nd ed. Paris, Office international des epizooties. 54 pp. (Available in English, French and Spanish)

Tanne, E. 1980. The use of ELISA for the detection of some nepoviruses in grapevines. *Proc. 7th Meet. ICVG,* Niagara Falls, NY, USA, 1980, p. 293-296.

Van Regenmortel, M.H.V. 1982. *Serology and immunochemistry of plant viruses*. New York, Academic Press. 302 pp.

Vela, C., Cambra, M., Cortes, E., Moreno, P., Miguet, S.G., Perez de San Roman, C. & Sanz, A. 1986. Production and characterization of monoclonal antibodies specific for citrus tristeza virus and their use in diagnosis. *J. Gen. Virol.,* 67: 91-96.

Walter, B. & Etienne, L. 1987. Detection of the grapevine fanleaf virus away from the period of vegetation. *J. Phytopathol.,* 120: 355-364.

Walter, B., Vuittenez, A. Kuszala, A., Stocky, G., Buckard, J. & Van Regenmortel, M.H.V. 1984. Détection sérologique du virus du court-noué dans la vigne par le test ELISA. *Agronomie,* 4: 527-534.

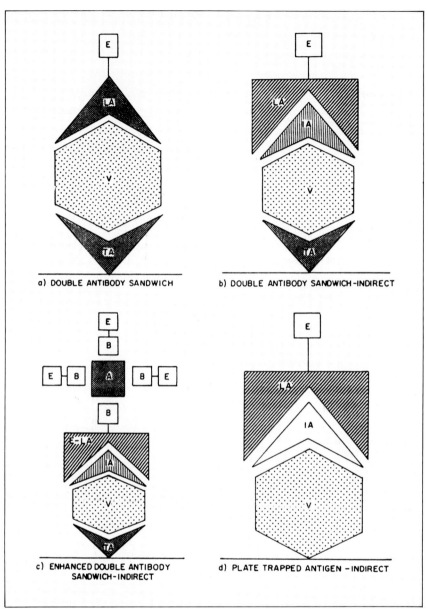

FIGURE 262

Diagram of the components of four popular types of ELISA: a) *Double antibody sandwich ELISA (DAS)*. The most widely used form of ELISA for plant pathogens. The wells of the ELISA plate (the solid phase or immunosorbent surface) are coated with an unlabelled antibody specific to the pathogen, which becomes the trapping antibody (TA). The antigen (V) is captured by the trapping antibody and detected by the enzyme-labelled antibodies (LA), which are normally from the same polyclonal antiserum used for trapping and detection ; b) *Double antibody sandwich indirect ELISA (DAS-I)*. The intermediate antibody (IA) is unlabelled and must be from a different animal species than the coating antibody. The LA is an antibody specific for the IA. If the F(ab')$_2$ antibody component is used for coating, the whole unlabelled antibody from the same animal can be used as the IA and is detected with protein A conjugated to an enzyme ; c) *Enhanced DAS-I*. This is similar to DAS-I, but the enzyme concentration on the LA is amplified by an additional treatment to increase sensitivity. Frequently, the LA is biotinylated to react to avidin-enzyme conjugates ; d) *Plate-trapped antigen indirect ELISA*. The antigen (V) is trapped directly on the plate surface and detected by using an unlabelled antibody specific to the antigen (IA) plus an enzyme-labelled antibody (LA) specific to the IA. Enhancement as shown for DAS-I is also possible

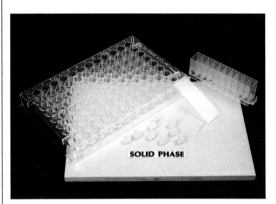

FIGURE 263
ELISA is normally done using plastic microtitre plates specially formulated for that purpose. Plates with 96 wells are most common, but other configurations exist. Plastic beads or strips can also be used as the solid phase (immunosorbent surface) and are convenient for small numbers of samples

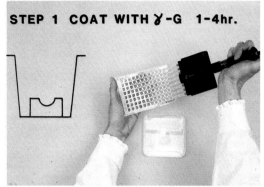

FIGURE 264
The first step in DAS or DAS-I ELISA is to coat the plate with the trapping antibody (TA in Figure 262). The trapping antibody is prepared at a concentration of 1 to 2 μg/ml in carbonate buffer and 100 to 200 μl are placed in each well (note use of an eight-channel pipette for fast loading) and incubated for 1 to 4 hours. Unbound antibody is removed by washing the plate with PBST (Figure 277)

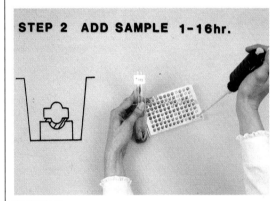

FIGURE 265
The second step in DAS and DAS-I is to load the test samples which have been previously prepared (see Figures 271 to 276), coded and arranged in sequence for easy loading. Each sample is placed in at least two wells, and each plate contains a positive and negative control sample for reference. Each plate is marked for identification and orientation. The plates are incubated for 2 to 4 hours at 30 to 37°C, or overnight at 4 to 6°C, and then washed with PBST to remove materials not bound specifically to the trapping antibody

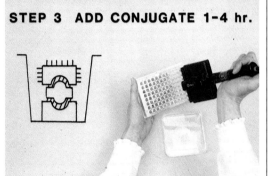

FIGURE 266
The third step in DAS is to add the enzyme-labelled conjugate diluted to a predetermined optimum concentration in PBST (in DAS-I this step is preceded by application of the intermediate antibody). Incubate 2 to 4 hours at 37°C. Wash at least three times with PBST to remove unbound conjugate from the wells. Thorough washing is important prior to substrate addition to avoid non-specific reactions

FIGURE 267
The fourth step is preparation and addition of the substrate solution to the test plate. Substrates should be freshly made. Colour change in the substrate will be proportional to the number of enzyme-labelled antibodies bound to the antigen present

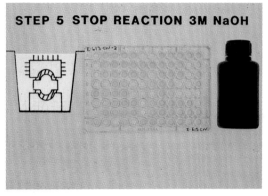

FIGURE 268
Once an appropriate level of reaction is reached, the reaction can be stopped by addition of 50 μl of 3 M NaOH and then the plate is evaluated. Plates may be read several times during the incubation period to determine the rate of reaction. Plates may also be frozen for future evaluation

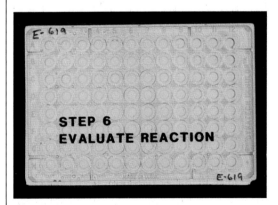

FIGURE 269
The reaction is evaluated by determining the degree of colour change in the substrate. This may be done visually by scoring the degree of reaction (e.g. negative, weak, moderate or strong). Comparisons are made with both the healthy and positive control samples

FIGURE 270
The degree of reaction may be determined by measuring the colour change photometrically (at 405 nm for alkaline phosphatase) by use of a plate reader, which measures and reports the absorbance in each well. By reading the plate at several timed intervals, the rate of reaction can also be calculated

FIGURE 271
Proper selection of tissues is important for successful ELISA. With nepoviruses and other sap-transmissible viruses, antigen concentration is usually highest in new flushes of growth (right), whereas with closteroviruses, aged leaves collected in autumn (left) represent a better antigen source than young leaves

FIGURE 272
Cortical tissues are excellent antigen sources for nepoviruses and phloem-limited viruses (e.g. closteroviruses and grapevine fleck virus). The figure shows (from left to right) the sequence of operations from peeling outer bark off mature canes to scraping cortex down to xylem. Cortical shavings (upper right) are extracted in buffer and used for ELISA. Leaf petioles (lower right) can also be used successfully for serological detection of closteroviruses and other pathogens inhabiting conducting tissues (e.g. Pierce's disease bacterium)

FIGURE 273
Sample tissue is finely diced to avoid clogging the homogenizer and is placed in the grinding vessel at a ratio of 1 part tissue to 10 to 20 parts buffer. Samples may be stored frozen in buffer if immediate testing is not convenient

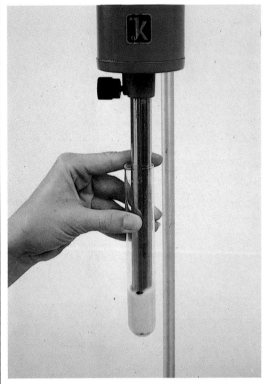

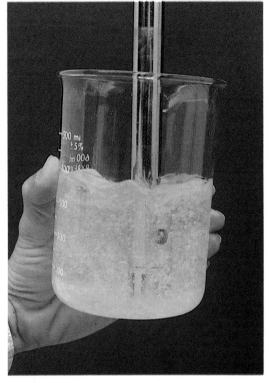

FIGURE 274
A dispersion homogenizer is very convenient for grinding tough fibrous tissue samples such as bark and leaf midrib tissue. Samples can also be ground by other methods including a pestle and mortar (adding an abrasive makes it easier)

FIGURE 275
The shaft of the homogenizer must be rinsed in clean water between samples to avoid contamination

FIGURE 276
Tissue extracts usually contain debris which makes
pipetting difficult. One way to avoid clogging the tip is to
pipette through a filter. The extract can also be centrifuged
or the tip of the pipette can be cut off to create a larger bore

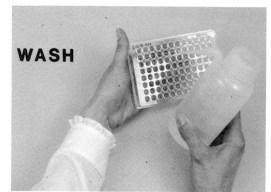

FIGURE 277
Washing is a very important part of ELISA. Each plate is
washed at least three times between each step. The plate is
emptied by rapidly inverting it over a sink, blotted on a
clean towel, filled with wash solution (normally PBST) by a
squeeze bottle or other dispenser and incubated with
gentle agitation for several minutes. The cycle is repeated.
The final wash solution may be left in the plate until the
next step is initiated

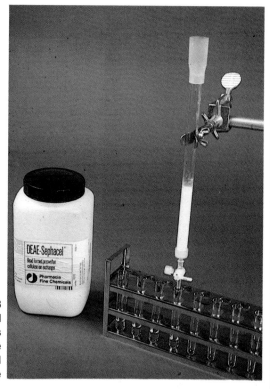

FIGURE 278
Antibodies are immunoglobulins (IgG) and can be purified
from whole antisera by use of commercially prepared kits
or by precipitation of the serum with ammonium sulphate
and chromatography of the dialysed precipitate on a small
column of DEAE cellulose prepared for that purpose

Plate No. _____
Date _____/_____/_____
Expt. No. _____

Plate No. _____ Conc._____ Time _____ Temp. _____
Coating Ab_____ Conc._____ Time _____ Temp. _____
Antigen _____ Conc._____ Time _____ Temp. _____
Second Ab _____ Conc._____ Time _____ Temp. _____
Conj. _____ Conc._____ Time _____ Temp. _____
Subst _____ Conc._____ Time _____ Temp. _____

NOTES. _____

	1	2	3	4	5	6	7	8	9	10	11	12
A												
B												
C												
D												
E												
F												
G												
H												

FIGURE 279
Data sheet for recording ELISA results

Immunosorbent electron microscopy (ISEM) and antibody coating

G.P. Martelli

The principle of immunosorbent electron microscopy (ISEM) is the selective trapping of plant viruses on to electron microscope grids precoated with a specific antiserum. This technique has been described in a number of papers and review articles (Derrick, 1973; Milne and Luisoni, 1977; Garnsey *et al.*, 1979; Roberts and Harrison, 1979; Russo, Martelli and Savino, 1980; Van Regenmortel, 1982; Milne and Lesemann, 1984; Hampton, Ball and De Boer, 1990) to which the reader is referred for comprehensive information.

ISEM may be combined with antibody coating (often referred to as "decoration"), a procedure whereby virus particles trapped on the microscope grid are exposed to the homologous antiserum, thus becoming visibly covered with antibody molecules.

The consensus is that ISEM is highly reliable (there are virtually no false positives), as sensitive as ELISA, fast (results can often be obtained within one or two hours) and operationally simple (it requires tools and reagents readily available in most laboratories).

Unfortunately, ISEM requires an electron microscope and accordingly is not suitable for large-scale routine testing. Specimens for ISEM, however, can readily be prepared in laboratories with no electron microscope facilities and can then be shipped for observation (even over long distances) to properly equipped institutions.

BASIC TOOLS AND REAGENTS

The following are required (see Figures 280 to 282):

- porcelain mortars 6 cm in diameter or smaller, or glass microscope slides and glass rods
- carborundum powder (600 mesh) or quartz sand
- bench centrifuge with appropriate glass or plastic conical tubes
- fine straight-point tweezers
- Petri dishes 9 cm in diameter
- bars of dental wax, silicone-treated paper, or parafilm
- Pasteur pipettes
- electron microscope grids (400 mesh) covered with carbon film
- protectants:
    2.5 to 5 percent aqueous solution of nicotine
    2 percent aqueous polyvinylpyrrolidone (PVP)
    1 percent aqueous polyethylene glycol (PEG), MW 6 000 to 7 500
- phosphate buffer 0.1 M, pH 7.0, made from the following stock solution (1 M):
    136.09 g of KH_2PO_4 in distilled water to 1 litre (solution A)
    268.077 g of Na_2HPO_4 in distilled water to 1 litre (solution B)
  Mix 3.86 ml of solution A with 6.14 ml of solution B and dilute tenfold

- distilled water
- staining solutions:

  1 to 2 percent uranyl acetate in distilled water, pH not adjusted

  2 percent sodium or potassium phosphotungstate in distilled water, adjusted to pH 7 with NaOH or KOH
- appropriate antiserum

PREPARATION OF TISSUE EXTRACTS

Extracts may be prepared from tissues of different organs of field- or greenhouse-grown plants (leaves, roots, bark, dormant or breaking buds) (Figure 283) or vectors (insects, nematodes). Plant tissues (usually 100 to 200 mg) are ground in a mortar in the presence of carborundum powder or quartz sand and 0.3 to 0.5 ml of phosphate buffer or, especially with grapevine and stone fruits, one of the above protectants (nicotine, PVP, PEG). When a smooth paste is obtained, 0.3 to 0.5 ml of buffer are added and the sample is ground again. The slurry is transferred to a centrifuge tube and centrifuged at 1 500 to 2 000g. The supernatant fluid is collected and used (Figure 284).

If a centrifuge is not available, tissue extracts can be further diluted with phosphate buffer to 1:15 to 1:20 with respect to tissue weight, and used as such.

Insect and nematode vectors are crushed with a glass rod on a glass slide in a droplet of buffer or protectant. A droplet of buffer is then added and the extract is ready to be used.

ANTISERUM DILUTIONS

The purpose of using an antiserum is twofold: to coat EM grids for trapping virus particles and to "decorate" virus particles by attachment of antibody molecules to the antigenic sites of the particles. Crude antisera are perfectly suitable for both uses, provided that they are properly diluted. For coating of grids, dilute antiserum to

near or above its end point (usually 1:1 000 to 1:5 000) with buffer. For decorating virus, dilute antiserum to 1:10 to 1:100 with buffer. Use of freshly diluted antisera is advisable.

PRECOATING OF EM GRIDS

In certain cases, precoating of EM grids with protein A, a bacterial wall protein that binds specifically to the basal part (Fc portion) of antibody molecules, can be advantageous. For instance, protein A allows trapping of more virus particles because of the richer antibody layer on the grid. It also allows the use of undiluted, low-titre (1:8 to 1:16) antisera which would not be suitable after high dilution as required by ordinary ISEM.

Protein A is diluted in phosphate buffer at a final concentration of 10 to 100 μg per ml, a drop is placed on the grid for five minutes at room temperature and the excess is rinsed off before exposure to antiserum.

ANTISERUM COATING OF EM GRIDS

Drops of diluted antiserum (1:1 000 to 1:5 000) are placed on dental wax or other hydrophobic supports (parafilm strips, silicone-treated paper) in a plastic Petri dish containing moist filter paper (moist chamber) (Figure 285). A freshly prepared carbon-coated grid is gently placed, film-down, on top of each antiserum drop and floated for 5 to 10 minutes at room temperature. Grids are then removed with tweezers and rinsed.

RINSING THE EM GRIDS

Throughout the ISEM procedure, grids must be carefully rinsed to obtain clean preparations. Buffer rinse (Figure 286) is used after protein A precoating, antiserum coating and incubation of the grid with tissue extract. Distilled water rinse is used after second antibody coating (decoration of virus particles), before negative staining, for uranyl acetate precipitates in the presence of phosphate ions, or at neutral pH.

Two rinsing procedures can be utilized:
- Grids are floated on drops of buffer or distilled water, as appropriate, for 5 to 10 minutes.
- Grids are retained in tweezers, held vertically and rinsed with 25 to 30 drops of buffer or water from a Pasteur pipette held close to the grid.

NEGATIVE STAINING

Negative stain can be applied with either system used for rinsing, i.e. floating grids on small drops of the staining solution for 30 seconds to one minute, or applying the stain dropwise (five drops) with a Pasteur pipette.

SUMMARY OF THE PROCEDURE

- Prepare tissue extracts; place drops of extract in a moist chamber on a hydrophobic support (Figure 284).
- Float antiserum-coated grids film-down, one on each drop of extract (Figure 284). Incubate at room temperature or in the cold (4°C) for 6 to 8 hours.
- Rinse with 25 to 30 drops of phosphate buffer from a Pasteur pipette (Figure 286). Drain with filter paper.
- Place drops of antiserum at dilution of 1:10 to 1:100 in the moist chamber on a hydrophobic support (Figure 287).
- Float grids on antiserum drops for 10 to 15 minutes at room temperature (Figure 287).
- Rinse with 25 to 30 drops of distilled water from a Pasteur pipette (Figure 288).
- Apply negative stain dropwise (five drops) from a Pasteur pipette (Figure 289). Remove excess with filter paper.
- Grids are ready for observation (Figure 290). Observe with the electron microscope and read the results.

REFERENCES

Derrick, K.S. 1973. Quantitative assay for plant viruses using serologically specific electron microscopy. *Virology*, 56: 652-653.

Garnsey, S.M., Christie, R.G., Derrick, K.S. & Bar-Joseph, M. 1979. Detection of citrus tristeza virus. II. Light and electron microscopy of inclusions and virus particles. *Proc. 8th Conf. IOCV*, p. 9-16. Riverside, CA, IOCV.

Hampton, R., Ball, E. & De Boer, S., eds. 1990. *Serological methods for detection and identification of viral and bacterial plant pathogens. A laboratory manual.* St Paul, MN, USA, Am. Phytopathol. Soc. Press. 389 pp.

Milne, R.G. & Lesemann, D.-E. 1984. Immunosorbent electron microscopy in plant virus studies. *Methods Virol.*, 8: 85-101.

Milne, R.G. & Luisoni, E. 1977. Rapid immune electron microscopy of virus preparations. *Methods Virol.*, 6: 265-281.

Roberts, I.M. & Harrison, B.D. 1979. Detection of potato leafroll and potato mop-top viruses by immunosorbent electron microscopy. *Ann. Appl. Biol.*, 93: 289-297.

Russo, M., Martelli, G.P. & Savino, V. 1980. Immunosorbent electron microscopy for detection of sap-transmissible viruses of grapevine. *Proc. 7th Meet. ICVG*, Niagara Falls, NY, USA, 1980, p. 251-257.

Van Regenmortel, M.H.V. 1982. *Serology and immunochemistry of plant viruses.* New York, Academic Press. 302 pp.

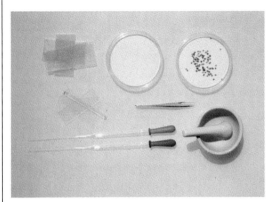

FIGURE 280
Basic tools for use with immune electron microscopy.
From left to right: dental wax bars, Petri dish, carbon-
coated electron microscope grids, glass rods and slides,
straight-point tweezers, Pasteur pipettes, porcelain mortar

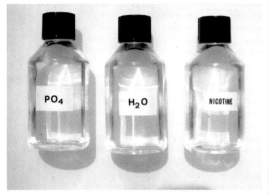

FIGURE 281
Extraction and rinsing media: phosphate buffer (PO_4),
distilled water (H_2O) and 2.5 percent aqueous nicotine

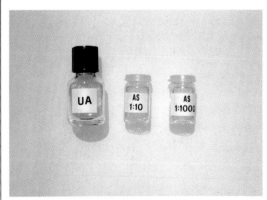

FIGURE 282
Uranyl acetate (1 to 2 percent solution in distilled water)
and tenfold and thousandfold dilutions of antiserum for
"decorating" and "trapping" virus particles on the EM grid,
respectively

FIGURE 283
Plant organs commonly used for preparation of extracts:
leaves, bark, roots, buds

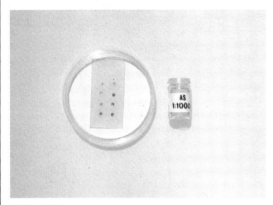

FIGURE 284
Drops of plant extract, obtained by grinding tissues on a mortar, on which antibody-coated EM grids are being floated for particle "trapping" (upper wax bar). Lower wax bar supports drops of phosphate buffer on which EM grids are being floated for rinsing

FIGURE 285
Petri dish with a dental wax bar on which EM grids are being floated on drops of a thousandfold dilution of antiserum for antibody coating

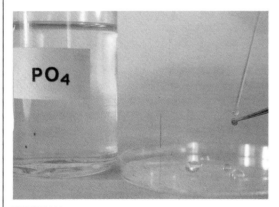

FIGURE 286
Rinsing EM grids with phosphate buffer applied dropwise

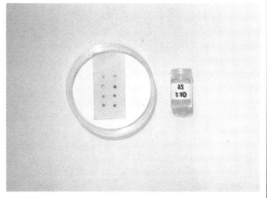

FIGURE 287
EM grids being floated on drops of tenfold diluted antiserum for particle "decoration"

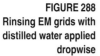

FIGURE 288
Rinsing EM grids with distilled water applied dropwise

FIGURE 289
Staining EM grids with uranyl acetate applied dropwise

FIGURE 290
Petri dish with EM grids ready for observation

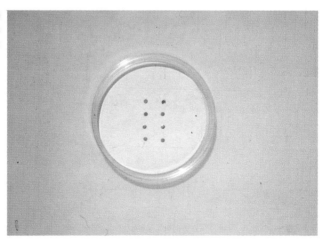

Detection and identification of viroids

J.S. Semancik

HOST PLANTS FOR PURIFICATION AND BIOASSAY

Indicator plants used for viroid purification and bioassay commonly respond with some form of stunting reaction which may be accompanied by leaf symptoms of rugosity, epinasty, mottling and chlorotic spotting and vein browning. However, since viroid replication may occur in the absence of any discernible symptoms, all inoculated species should be extracted and analysed for viroid content. Plant species suspected of containing viroids can be analysed directly, provided that the extraction conditions necessary to obtain a quality nucleic acid preparation are determined.

In viroid transmission studies, the preferred host plants are seedlings, which in most cases are viroid-free, or vegetatively propagated plant sources made viroid-free by shoot-tip culturing. Woody plant species, such as citron and grapevine, can be inoculated by slashing the stem with a razor-blade moistened with inoculum. With more succulent species, such as tomato, a needle or fine Pasteur pipette is used to puncture the hypocotyl of a very young plant at the point where a drop of inoculum has been applied.

Major indicator hosts

- Chrysanthemum (*Chrysanthemum morifolium* cv. Bonnie Jean) (see Brierley, 1953)
- Citron (*Citrus medica* cv. Etrog) (see Calavan *et al.*, 1964)
- Cucumber (*Cucumis sativus* cv. Suyo) (see Sasaki and Shikata, 1977; Van Dorst and Peters, 1974)
- Gynura (*Gynura aurantiaca*) (see Weathers and Greer, 1968)
- Petunia (*Petunia hybrida*) (see Weathers *et al.*, 1967)
- Tomato (*Lycopersicon esculentum* cv. Rutgers) (see Raymer, O'Brien and Merriam, 1964)

TISSUE EXTRACTION AND PURIFICATION

Ultimate success in detecting viroids as discrete bands on polyacrylamide gels is dependent upon the quality of the nucleic acid preparations obtained from infected tissues. High concentrations of phenolic and acidic compounds can seriously interfere with the recovery of all nucleic acid species. Therefore, the composition of the extraction medium must be customized to the particular tissue under investigation to assure the consistency of factors such as a pH maintained at about 6.5 to 9.0, the presence of appropriate additives such as polyvinylpyrrolidone to neutralize the effects of polyphenols, and adequate concentration of anti-oxidants.

Two protocols commonly used for the extraction of a tissue are presented. The procedures differ basically in the treatment of the aqueous phase from the initial phenol-extraction step (Figure 291). Concentration by ethanol precipitation is employed with tissue extracts from plants such as citron and tomato from which good nucleic acid preparations can routinely be recovered. "Trapping" of nucleic acids, including viroid RNA, on CF-11 cellulose (Figure 292) has been customized for use with direct extraction of grapevine tissue or other tissues from which nucleic acids are difficult to recover.

Even though the primary focus of the procedures presented here is the analysis of citrus, the alternative approach indicated for grapevines should be employed if nucleic acid preparations are either difficult to obtain or of poor quality. The designation of the procedures for grapevines simply indicates the plant tissues for which the technique was developed and does not imply an exclusive application. To date, grapevine tissues have been the most challenging for recovery of nucleic acid preparations of high quality and adequate quantity for analysis of viroid content. Therefore, the information developed for this tissue may become valuable for the analysis of other species.

Unless one can demonstrate the recovery of a typical profile of host nucleic acids, it is difficult to evaluate the relative concentration or even the very presence of viroid molecules. Therefore, it is good practice to inspect the presence and relative concentration of particularly the 4S and 5S RNA components of 2 M LiCl soluble nucleic acids following electrophoresis in the native polyacrylamide gels.

Materials (Figures 292 to 297)

- Infected tissue: fresh tip tissue which is actively growing and collected at least two to six weeks post-inoculation of herbaceous hosts and two to six months post-inoculation of woody species is preferred. If tissue is to be collected and stored for extraction at a later time, it should be powdered in liquid nitrogen and held at -20°C.
- Extraction medium (EM-1) for citrus species and herbaceous plants:

    Buffer (0.4 M Tris-HCl, pH 8.9)
    SDS (sodium dodecyl sulphate), 1 percent
    EDTA (ethylenedinitrilotetraacetate), 5 mM, pH 7.0
    MCE (mercaptoethanol), 4 percent
- Extraction medium (EM-2) for grapevines and plants containing a high concentration of phenols and acidic compounds:

    Buffer (0.5 M Na_2SO_3)
    SDS, 1 percent
- Resuspension medium (RM), TKM buffer:

    Tris, 10 mM
    KCl, 10 mM
    $MgCl_2$, 0.1 mM
    Adjust to pH 7.4 with HCl
- LiCl, 4 M
- PVP (polyvinylpyrrolidone), 20 percent (4X stock)
- Sodium acetate, 3 M, pH 5.5
- Phenol (water-saturated) adjusted to pH 7 with 1 N NaOH
- Ethanol, 95 to 100 percent
- CF-11 fibrous cellulose powder (Whatman)
- STE buffer 10X stock (1.0 M NaCl, 0.50 M Tris-HCl, pH 7.2, 10 mM EDTA)
- Dialysis tubing
- Homogenizer: Virtis, high speed (50 000 rpm)(Figure 299); food blender, low speed (15 000 rpm) (*Note:* Yield of viroid can be significantly affected by the method of homogenization)
- Centrifuge, low speed (5 000 to 10 000 rpm or 6 000 to 12 000g), refrigerated (Figure 298)
- Magnetic stirrer at 4°C

Extraction of citrus and most herbaceous species

1. Grind tissue in 1 ml pre-cooled EM-1 and 3 ml phenol per gram tissue, in an ice bath if possible.

Note: Phenol can cause severe burns; therefore, protection for hands (disposable gloves) as well as a plastic barrier should be employed to guard against accidents during homogenization.

2. Transfer to centrifuge tubes or bottles and centrifuge for 20 minutes at 7 000 to 12 000g.

3. A clear but pigmented aqueous layer is found over a solid interface of plant debris and a lower heavily pigmented phenol phase. The plant debris may also form a pellet below a liquid bilayer between the aqueous and phenol phases.

4. Remove aqueous layer and add a 1/10 volume of 3 M sodium acetate, pH 5.5 and a minimum of three volumes of 95 to 100 percent ethanol. Hold at -20°C for 30 minutes or an indefinite period.

5. Centrifuge for 20 minutes at 7 000 to 12 000 g.

6. Discard supernatant and drain pellets containing nucleic acids until excess ethanol is removed.

7. Cover pellets with a minimum volume of RM [1 to 10 ml per 5 to 100 g tissue, fresh weight (FW)] and resuspend with agitation.

8. Transfer slurry to dialysis tubing.

Note: At this point the solution may appear quite turbid and particulate, but it should clear considerably with dialysis.

9. Dialyse with rapid stirring on a magnetic stirrer at 4°C overnight against 1 litre of RM.

10. Remove sample from dialysis tubing to centrifuge tubes and add one volume of 4 M LiCl. Hold at 4°C for 4 hours or overnight.

11. Centrifuge for 20 minutes at 7 000 to 12 000g.

12. Retain supernatant containing LiCl-soluble nucleic acids (mainly DNA, 4S and 5S RNA, dsRNAs and viroids). Discard the pellet of LiCl-insoluble nucleic acids (mainly ribosomal RNA).

13. Add a minimum of three volumes of 95 to 100 percent ethanol and hold at -20°C for 30 minutes or overnight.

14. Centrifuge for 20 minutes at 7 000 to 12 000g.

15. Decant and drain ethanol from pellets and dry *in vacuo*.

16. Resuspend pellets in an appropriate volume of RM [100 µl per 5 g tissue (FW)].

17. Store at -20 to -80°C.

Note: These preparations are sufficiently purified for routine viroid detection procedures and infectivity tests. However, the quality of the analysis will be markedly improved by further processing by cellulose chromatography.

Extraction of grapevines and tissues from which it is difficult to recover nucleic acids

1. Grind tissue as indicated above, substituting EM-2.

2. Centrifuge as above.

3. Note as above.

4. Remove aqueous phase from above interface and lower phenol layer.

5. Make solution to 35 percent ethanol and 1X STE with stirring. Add dry CF-11 cellulose powder [1 g per 5 g (FW) tissue extract]. Stir for 2 hours or overnight at room temperature.

6. Collect cellulose by centrifuging at 7 000g for 10 minutes.

7. Discard supernatant and wash cellulose pellet with a solution of 30 percent ethanol in 1X STE buffer with agitation (Figure 300).

8. Collect cellulose as in Step 6.

9. Repeat washing procedure with 30 percent ethanol-STE solution two or three times until all traces of pigmented materials have been removed from the wash solution.

10. With the cellulose in 30 percent ethanol-STE, form a chromatography column and continue to wash cellulose with three to four void volumes of 30 percent ethanol-STE.

11. Elute bound nucleic acids using two to three void volumes of STE buffer, collecting the eluent in a serial manner and not as a single batch.

12. Add 10 percent volume of 3 M sodium acetate, pH 5.5, and a minimum of three volumes of ethanol. Let solution stand at -20°C for 30 minutes or longer as convenient.

13. Collect precipitated nucleic acids by centrifugation at 12 000g for 20 minutes.

14. Discard supernatant and allow pellet to drain until reasonably dry.

15. Resuspend pellet in minimum amount of RM buffer.

16. Add one volume of 4 M LiCl and let stand at 4°C for 4 hours or overnight.

17. Centrifuge at 12 000g for 20 minutes. Retain supernatant of 2 M LiCl-soluble nucleic acids.

18. Add a minimum of three volumes of ethanol to the supernatant and let stand at -20°C for 30 minutes or longer.

19. Collect precipitated nucleic acids by centrifugation at 12 000g for 20 minutes.

20. Discard supernatant, drain liquid from pellet and dry *in vacuo*.

21. Resuspend pellet in a minimum volume of RM buffer, usually in the range of 100 µl per 5 to 10 g tissue (FW).

22. Store samples at -20 to -80°C prior to analysis by polyacrylamide gel electrophoresis or infectivity.

See Duran-Vila, Flores and Semancik, 1986; Semancik *et al.*, 1975; Semancik, Rivera-Bustamante and Goheen, 1987.

CF-11 CELLULOSE CHROMATOGRAPHY

This technique (Figures 301 and 303) can be utilized routinely as a preparative procedure for the removal of DNA and other pigmented components of viroid-containing LiCl-soluble nucleic acid preparations. The property of selective binding of viroid RNA at specific ethanol concentrations can also be exploited in recovering viroids from tissue extracts, as was demonstrated in the "trapping" procedure presented in the previous section.

More recently, an analytical approach to CF-11 cellulose chromatography has been introduced (Semancik, 1986) to characterize different viroid RNAs by serial elution with an ethanol gradient. This procedure can be utilized to remove contaminating host RNAs from viroid preparations as well as to separate individual viroids with selective elution by different ethanol concentrations.

Materials
- CF-11 cellulose powder, fibrous (Whatman)
- STE buffer (0.1 M NaCl, 1 mM EDTA, 0.05 M Tris-HCl, pH 7.2)
- 95 to 100 percent ethanol
- syringe barrels (disposable) or chromatography columns
- GF/C glass microfibre filter discs (Whatman 2.4 cm)

Preparative chromatography

1. An aqueous sample containing nucleic acids is made to 35 percent ethanol in STE buffer.

2. Apply the solution to a CF-11 cellulose column which has been equilibrated with 35 percent ethanol-STE.

Note: (a) The amount of cellulose used is dependent upon the amount of nucleic acid in the preparation. A proportion of 1 to 10 g cellulose per 5 to 100 g (FW) extraction is usually adequate. (b) A "trapping" procedure can also be employed with dry cellulose added directly to the aqueous phase from a phenol extraction made to 35 percent ethanol-STE.

3. Wash the cellulose with sufficient 30 percent ethanol-STE to remove all traces of colour from the cellulose or with a volume equivalent to at least four to six column void volumes.

4. Elute the nucleic acids retained on the column with two to four void volume equivalents of STE (0 percent ethanol).

5. Precipitate nucleic acids with addition of 1/10 volume of 3 M sodium acetate, pH 5.5, and at least three volumes of 95 to 100 percent ethanol, and hold at -20°C for 30 minutes or longer.

6. Centrifuge for 20 minutes at 12 000g.

7. Dry pellet *in vacuo* and resuspend in TKM buffer (resuspension medium from extraction procedure).

Analytical chromatography

1. Nucleic acid sample from the extraction procedure or preferably a sample pre-treated on a preparative CF-11 column is made to 35 percent ethanol-STE.
2. Apply to a chromatography column containing an adequate amount of CF-11 cellulose equilibrated with 35 percent ethanol-STE.
3. Wash column with 35 percent ethanol-STE (four to six column void volumes).
4. Elute with 25 percent ethanol-STE, collecting two to four column void volumes. Retain eluent.
5. Wash column with 25 percent ethanol-STE (four to six column void volumes).
6. Elute with 20 percent ethanol-STE, collecting two to four column void volumes. Retain eluent.
7. Continue alternating wash and elution cycles either with a progressively reduced ethanol concentration (reduced in 5 percent increments, for example) or with a decreasing linear ethanol gradient to a final elution in STE buffer.
8. Precipitate nucleic acids with addition of 1/10 volume of 3 M sodium acetate, pH 5.5, and at least three volumes of 95 to 100 percent ethanol, and hold at -20°C for 30 minutes or longer.
9. Centrifuge for 20 minutes at 12 000g.
10. Dry pellet *in vacuo* and resuspend in RM buffer.
11. Analyse by sequential PAGE under native and denaturing conditions.

See Barber, 1966; Duran-Vila, Flores and Semancik, 1986; Franklin, 1966; Semancik, 1986.

POLYACRYLAMIDE GEL ELECTROPHORESIS (PAGE)

Optimum resolution of viroid RNA is obtained by a sequential gel electrophoresis procedure (Figure 302) involving migration of the sample into a standard gel (5 percent PAGE) (Figure 304), followed by excision of a piece of the gel, which is then placed in contact with a second, denaturing gel (dPAGE) containing 8 M urea (Figure 305). This procedure exploits the unique properties of the single-stranded closed circular structure of the viroid for the separation of a distinct band. Placement of the excised gel piece in contact with the top (Semancik and Harper, 1984) or the bottom (Schumacher, Randles and Riesner, 1983) of the denaturing gel with migration to the anode will produce similar results.

The discontinuous pH dPAGE (Rivera-Bustamante, Gin and Semancik, 1986) with the gel cast at pH 6.5 (TAE buffer) but migrated in a pH 8.3 running buffer (TBE buffer) enhances the separation between the circular and linear molecular forms of the viroid. In addition, the background of host nucleic acids is reduced, which aids in the recovery of pure viroid preparations for physical characterization and hybridization analysis.

Verification of a suspected viroid can be made by a PAGE analysis sequence involving:

- non-denaturing 5 percent PAGE, with excision of a strip of the gel in the "viroid zone" as defined by citrus exocortis viroid (CEV) and avocado sun blotch viroid (ASV, ASBV);
- dPAGE (pH 6.5), with excision of any slowly migrating band suspected of being a viroid circular form;
- dPAGE (pH 8.3), with resolution of two distinct bands containing the viroid circular form and the linear molecular form, generated from circles during electro-phoresis.

Staining with ethidium bromide to visualize nucleic acid bands is necessary when gels are to be subjected to a second electrophoresis and/or when biologically active viroid is to be recovered. Sensitivity of detection can be increased with silver nitrate staining. However, this procedure renders the viroid inactivated and immobilized in the gel.

Materials

- Electrophoresis chamber and casting apparatus: glass plates, spacers, clamps, sample-well comb (apparatus of various sizes are available through commercial sources or can easily be custom fabricated)
- Power supply (100 mA, 1 000 volt or greater capacity)
- Ultraviolet transilluminator
- Polaroid camera
- Stock solutions for 5 percent gels:

Stock A:

Acrylamide, 30 g

Bisacrylamide, 0.75 g

Dissolve in distilled water, bring to 100 ml and filter.

Stock B:

Bring 2 ml tetramethylethylenediamine (TEMED) to 100 ml with distilled water.

Stock C:

Tris, 120 mM

Sodium acetate·3H$_2$O, 60 mM

Sodium EDTA, 3 mM

Dissolve in distilled water, adjust to pH 7.2 with glacial acetic acid.

Note: This solution is equal to 0.3X Stock D; therefore it can also be made by diluting 30 ml of Stock D to 100 ml with distilled water.

Stock D:

TAE buffer, pH 7.2, 10X

Tris, 400 mM

Sodium acetate·3H$_2$O, 200 mM

Sodium EDTA, 10 mM

Dissolve in distilled water, adjust to pH 7.2 with glacial acetic acid.

Note: Since Stock D is a 10X concentration, it should be diluted before use as a running buffer.

Stock E:

Dissolve 2.5 g ammonium persulphate (10 percent) in 25 ml H$_2$O. Prepare fresh weekly.

Stock F:

TAE buffer, pH 6.5

Tris, 120 mM

Sodium acetate·3H$_2$O, 60 mM

Sodium EDTA, 3 mM

Dissolve in distilled water, adjust to pH 6.5 with glacial acetic acid.

Stock G:

Denaturing gel buffer (TBE, pH 8.3, 10X)

Tris, 225 mM

Boric acid, 225 mM

Sodium EDTA, 5 mM

Dissolve in distilled water. No adjustment of pH should be necessary. Dilute tenfold for use as a running buffer (TBE, pH 8.3, 1X).

- Urea
- Glycerol (60 percent)
- Migration tracking dyes:

Bromophenol blue, 0.3 percent in 60 percent glycerol

Xylene cyanol, 0.3 percent in 60 percent glycerol

- Ethidium bromide stock staining solution (5 mg per ml) (30 µl per 200 ml H$_2$O for staining gels)
- Silver staining solutions:

Ethanol (50 percent) + acetic acid (10 percent)

Ethanol (10 percent) + acetic acid (1 percent)

Silver nitrate (12 mM)

Potassium hydroxide (0.75 M) + formaldehyde (0.28 percent)

Sodium carbonate (0.07 M)

Native 5 percent PAGE

1. Assemble glass form to receive polymerization solution.

2. Mix contents of two beakers containing the following solutions in the indicated amounts or similar proportions:

Beaker 1

    12.0 ml distilled water

    10.0 ml Stock C

    2.4 ml Stock B

Beaker 2

    5.0 ml Stock A

    0.48 ml ammonium persulphate

3. Fill form, place sample-well comb and let stand for 30 minutes.

4. Withdraw sample-well comb and lower spacer. Attach to chamber and fill electrode reservoirs with 1/10 dilution of Stock D.

5. Mix samples with about 1/4 volume of glycerol and load into wells with fine-tip Pasteur pipettes. Load outermost wells with mixture of tracking dyes.

6. Apply constant current, 54 mA, at 4°C for 2.5 to 3 hours or until bromophenol blue dye has migrated about 8 cm and xylene cyanol has reached about 4 cm.

7. Remove gel from the chamber and form. Soak with gentle agitation in the ethidium bromide staining solution for 10 minutes.

8. View the gel directly over a UV trans-illumination source. Cut horizontal strip as defined by "viroid zone" (CEV-ASV) or smaller, depending upon viroid, and transfer to denaturing gel.

Denaturing PAGE (pH 6.5)

1. Assemble glass form for polymerization solution.

2. Prepare two beakers with the following contents:

Beaker 1

    14.4 g urea

    7.0 ml H_2O

    3.0 ml Stock F (TAE, pH 6.5)

    5.0 ml Stock A

Beaker 2

    2.5 ml Stock B

    0.5 ml ammonium persulphate

Dissolve the contents of Beaker 1 over low heat. After the urea is dissolved, rapidly mix the contents of the two beakers.

3. Immediately fill form, leaving a flat surface with sufficient space for the excised native gel piece. Allow to stand for a minimum of 1 hour.

4. Remove lower spacer and attach to chamber. Do not add buffer or any liquid to the gel surface until immediately prior to use.

5. After section has been removed from native gel, fill electrode reservoirs and cover top surface of gel with Stock G diluted tenfold (TBE buffer, pH 8.3, 1X).

6. Float excised section on to the top of the denaturing gel, making as close a contact as possible.

7. Add a few drops of xylene cyanol-glycerol mix next to the outer edges of the gel strip.

8. Apply constant current, 15 mA, at 24°C for about 4 hours or until the xylene cyanol tracking has migrated to within 0.5 cm of the bottom of the gel.

9. Remove gel from form and stain either with ethidium bromide for additional dPAGE (pH 8.3 gel) to confirm circular and linear forms or for elution of viroid bands for infectivity or for use as templates for cDNA probes; or with silver nitrate for maximum sensitivity of detection.

Denaturing PAGE (pH 8.3)

1. Follow the same set-up and running procedure as presented above.

2. Mix rapidly the contents of two beakers containing:

Beaker 1

    14.4 g urea

7.0 ml H$_2$O

3.0 ml Stock G (TBE buffer, pH 8.3, 10X)

5.0 ml Stock A

dissolved on low heat

Beaker 2

2.5 ml Stock B

0.5 ml ammonium persulphate

3. Stain completed gel as before with either ethidium bromide or silver nitrate.

Silver staining

1. Gel can be stained with silver directly or following ethidium bromide staining without additional treatment.

2. Soak gel at room temperature in solution of 50 percent ethanol + 10 percent acetic acid for at least 1 hour with gentle shaking. Overnight soaking can sometimes improve the background.

3. Soak gel at room temperature in solution of 10 percent ethanol + 1 percent acetic acid for 1 hour with gentle shaking.

4. Soak in solution of 12 mM AgNO$_3$ for 1 hour with gentle shaking.

5. Rinse thoroughly (three times) with distilled water.

6. Rinse rapidly with small volume of developer solution (0.75 M KOH + 0.28 percent HCHO) and discard solution.

7. Add fresh developer solution (100 to 200 ml) and observe until bands appear, usually within 20 minutes.

8. Add excess distilled water and allow gel to expand. This process reduces the background and improves the quality of photographs.

9. Developing reaction can be stopped with 0.07 M Na$_2$CO$_3$.

10. Photograph gel over a light-box using Polaroid film (Figures 306 to 309).

See Igloi, 1983; Morris and Wright, 1975; Rivera-Bustamante, Gin and Semancik, 1986; Schumacher, Randles and Riesner, 1983; Semancik and Harper, 1984.

INFECTIVITY OF NUCLEIC ACID FRACTIONS AND VIROID MOLECULES

The infectivity of a viroid-containing sample can be influenced by the quality of the preparation. In many cases, viroid transmission by highly purified preparations can be more difficult than by a more complex, less purified preparation, perhaps in part because the presence of host nucleic acids may protect the viroid molecule from inactivation. Therefore, a sample such as a 2 M LiCl-soluble fraction may be valuable to demonstrate the transmission properties and host range of suspected viroid-containing preparations.

Nevertheless, an essential proof for the detection of a viroid is the transmissibility of the putative viroid-like molecule. This can be provided by the recovery of the unique, transmissible viroid structure, the single-stranded circular RNA molecule, in highly purified form followed by transmission to a host plant.

Electro-elution of the circular forms of viroids, as detected in denaturing PAGE by ethidium bromide staining, has proved to be a highly efficient procedure for the recovery of biologically active, pure viroid.

Materials

- PAGE gel piece containing the viroid
- Electro-elution buffer (EB) (1/50 dilution of Stock D in PAGE procedure):

    8.0 mM Tris

    4.0 mM sodium acetate

    0.2 mM EDTA

    adjusted to pH 7.2 with acetic acid

- Electro-elution apparatus: A chamber can be constructed to accommodate a piece of dialysis tubing filled with EB into which the gel piece has been introduced. When placed in an electrical field the viroid will migrate from the gel but be retained in the liquid phase inside the tubing. Alternatively, a commercial apparatus (Unidirectional Electroelutor Model UEA) that does not require the dialysis tubing containment procedure is available from International Biotechnologies, Inc., PO Box 9558, New Haven, CT 06535, USA.

- Power supply (25 mA, 250 V)
- 3 M sodium acetate, pH 5.5
- Ethanol

Electro-elution

1. Prepare gel piece to be eluted within dialysis tubing or according to IBI instructions.
2. Apply about 125 V constant voltage for 30 minutes at room temperature.
3. Withdraw buffer sample containing eluted viroid, add 1/10 volume of 3 M sodium acetate, pH 5.5, plus at least three volumes of ethanol and hold at -20°C for 30 minutes or longer.

Note: The gel piece can be checked for incomplete elution of viroid by restaining with ethidium bromide and viewing over a UV transilluminator. If viroid still remains in the gel piece, the elution procedure can be repeated.

4. Centrifuge sample at 12 000g for 20 minutes. Pellets may be extremely small or invisible. Nevertheless, sufficient viroid to be detected by PAGE and silver staining or infectivity can be recovered many times.
5. Dry decanted centrifuge tubes *in vacuo* and resuspend pellets in appropriate volume of TKM buffer (RM) or desired medium.

REFERENCES

Barber, R. 1966. The chromatographic separation of ribonucleic acids. *Biochim. Biophys. Acta*, 114: 422-424.

Brierley, P. 1953. Some experimental hosts of the chrysanthemum stunt virus. *Plant Dis. Rep.*, 37: 343-345.

Calavan, E.C., Frolich, E.F., Carpenter, J.B., Roistacher, C.N. & Christiansen, D.W. 1964. Rapid indexing for exocortis of citrus. *Phytopathology*, 54: 1359-1362.

Duran-Vila, N., Flores, R. & Semancik, J.S. 1986. Characterization of viroid-like RNAs associated with the citrus exocortis syndrome. *Virology,* 150: 75-84.

Franklin, R. 1966. Purification and properties of the replicative intermediate of the RNA bacteriophage R17. *Proc. Natl. Acad. Sci. USA*, 55: 1504-1511.

Igloi, G. 1983. A silver stain for the detection of nanogram amounts of tRNA following two-dimensional electrophoresis. *Anal. Biochem.*, 134: 184-188.

Morris, T.J. & Wright, N.S. 1975. Detection on polyacrylamide gel of a diagnostic nucleic acid from tissue infected with potato spindle tuber viroid. *Am. Potato J.*, 52: 57-63.

Raymer, W.B., O'Brien, M.J. & Merriam, D. 1964. Tomato as a source of and indicator plant for the potato spindle tuber virus. *Am. Potato J.*, 41: 311-314.

Rivera-Bustamante, R. F., Gin, R. & Semancik, J.S. 1986. Enhanced resolution of circular and linear molecular forms of viroid and viroid-like RNA by electrophoresis in a discontinuous-pH system. *Anal. Biochem.*, 156: 91-95.

Sasaki, M. & Shikata, E. 1977. Studies on the host range of the hop stunt disease in Japan. *Proc. Jpn. Acad.*, 55: 103-108.

Schumacher, J. , Randles, J.W. & Riesner, D. 1983. A two-dimensional electrophoretic technique for the detection of circular viroids and virusoids. *Anal. Biochem.*, 135: 288-295.

Semancik, J.S. 1986. Separation of viroid RNAs by cellulose chromatography indicating conformational distinctions. *Virology*, 155: 39-45.

Semancik, J.S. & Harper, K.L. 1984. Optimal conditions for cell-free synthesis of citrus exocortis viroid and the question of specificity of

RNA polymerase activity. *Proc. Natl. Acad. Sci. USA,* 81: 4429-4433.

Semancik, J.S., Morris, T.J., Weathers, L.G., Rodorf, B.F. & Kearns, D.R. 1975. Physical properties of a minimal infectious RNA (viroid) associated with the exocortis disease. *Virology,* 63: 160-167.

Semancik, J.S., Rivera-Bustamante, R. & Goheen, A.C. 1987. Widespread occurrence of viroid-like RNA in grapevine. *Am. J. Enol. Vitic.,* 38: 35-40.

Van Dorst, H.J.M. & Peters, D. 1974. Some biological observations on pale fruit, a viroid-incited disease of cucumber. *Neth. J. Plant Pathol.,* 80: 85-96.

Weathers, L.G. & Greer, F.C. Jr. 1968. Additional herbaceous hosts of the exocortis virus of citrus. Abstract of a paper accepted for presentation at the Sixtieth Annual Meeting of the American Phytopathological Society, Columbus, Ohio, 2-6 September 1968. *Phytopathology,* 58: 1071.

Weathers, L.G., Greer, F.C. Jr & Harjung, M.K. 1967. Transmission of exocortis virus of citrus to herbaceous plants. *Plant Dis. Rep.,* 51: 868-871.

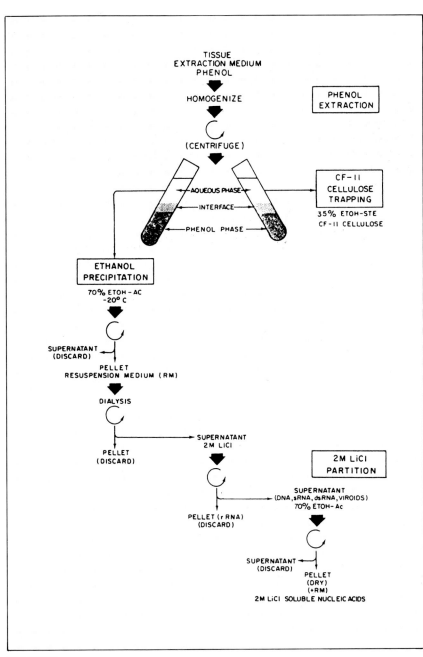

FIGURE 291
The procedure for tissue extraction and purification

FIGURE 292
Selection of citrus tissue for extraction

FIGURE 293
Selection of grapevine tissue for extraction

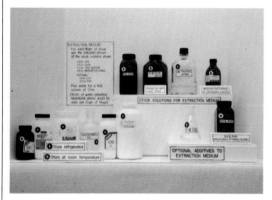

FIGURE 294
Components of extraction medium used for citrus

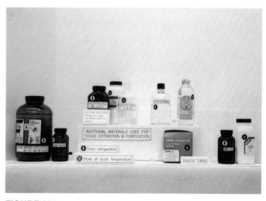

FIGURE 295
Components used in purification and concentration of nucleic acids

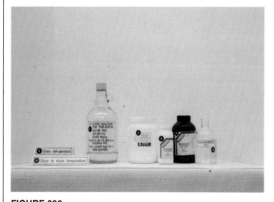

FIGURE 296
Components for resuspension of nucleic acid pellets

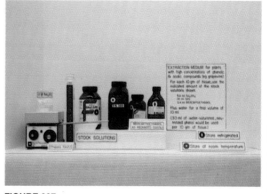

FIGURE 297
Components of extraction medium used for grapevines

FIGURE 298
Low-speed refrigerated centrifuge

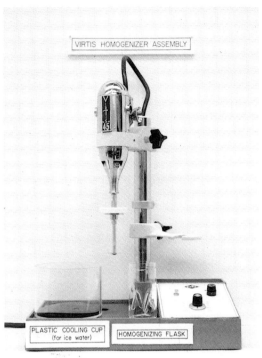

FIGURE 299
Virtis high-speed homogenizer

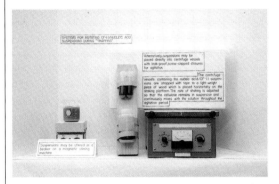

FIGURE 300
Apparatus used to agitate CF-11 cellulose for "trapping" of
nucleic acids

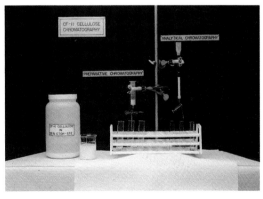

FIGURE 301
Apparatus used for preparative or analytical cellulose
chromatography

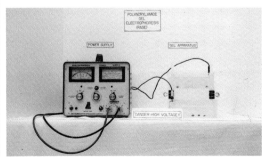

FIGURE 302
Apparatus used for polyacrylamide gel electrophoresis
(PAGE) and denaturing PAGE (dPAGE)

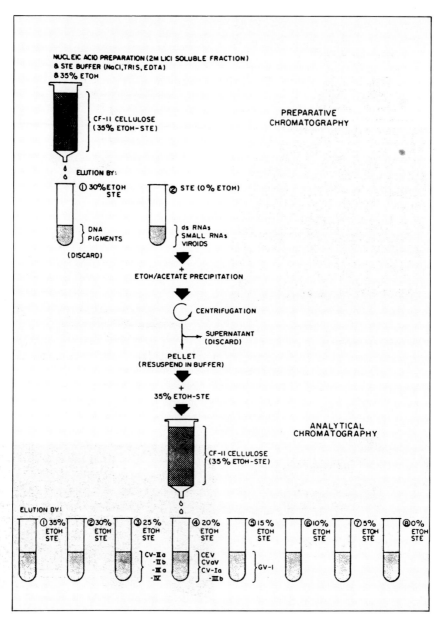

FIGURE 303
The procedure for CF-11 cellulose chromatography

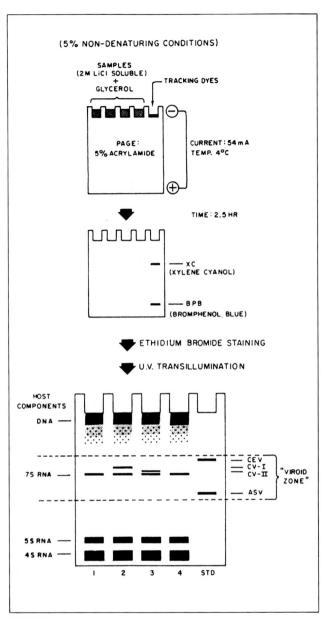

FIGURE 304
The procedure for polyacrylamide gel electrophoresis (PAGE)

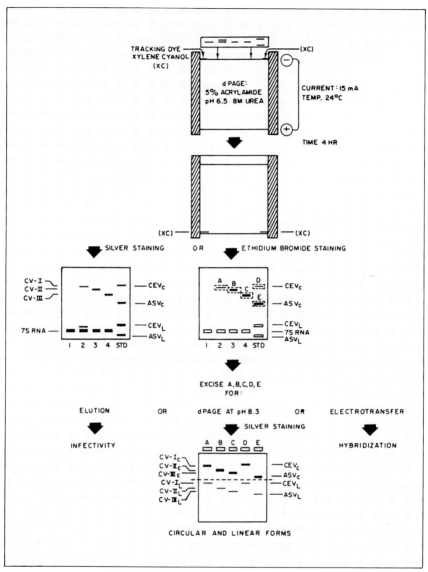

FIGURE 305
The procedure for denaturing polyacrylamide gel electrophoresis (dPAGE)

FIGURE 306
Polaroid photography apparatus
and transilluminator with
ultraviolet light source for
visualizing nucleic acid bands
stained with ethidium bromide

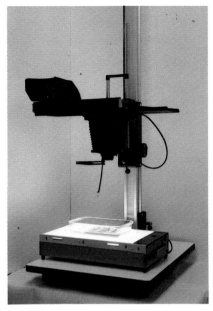

FIGURE 307
Polaroid photography apparatus and visible
light source for observing nucleic acid bands
stained with silver

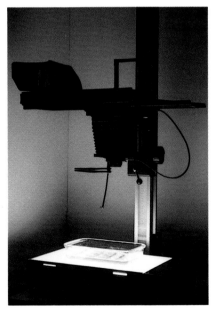

FIGURE 308
As for Figure 307 (with room lights off)

FIGURE 309
Silver-stained denaturing polyacrylamide gel
as seen through the Polaroid viewer

ELECTROPHORESIS

Isolation and analysis of double-stranded RNAs

D. Boscia

Double-stranded RNAs (dsRNAs) are paired molecules of viral genomic or subgenomic RNAs made up of a positive sense (+) RNA strand bound to a complementary negative sense (-) strand. dsRNAs represent replicative forms of viral RNAs which are formed during infection and may accumulate in the cells of diseased plants. They are stable molecule complexes with a size double that of genomic and subgenomic single-stranded RNAs and can be extracted from infected tissues. Their detection in plant extracts is regarded as a reliable indication of viral infection. By estimating the relative size of dsRNAs, it is possible to identify, with a fair approximation, the taxonomic group to which the eliciting virus belongs. However, specific identification of individual viruses is not possible. With grapevines, the study of dsRNA patterns is a useful complementary technique for determining the occurrence of non-mechanically transmissible, phloem-limited viruses.

Operations and materials for electrophoresis are the same as described for detection of viroids in the preceding chapter.

BUFFERS

- STE (0.1 M NaCl, 0.05 M Tris, 0.001 M EDTA, pH 6.8):
Prepare a stock solution 10X:
  Tris, 61 g
  NaCl, 58 g
  Na_2EDTA, 3.7 g

Adjust pH to 6.8 with HCl and make up to 1 litre with distilled water.
- STE, 16.5 percent ethanol:
  10X STE, 100 ml
  95 percent ethanol, 174 ml
Make up to 1 litre with distilled water.
- Bentonite
Prepare according to Fraenkel-Conrat, Singer and Tsugita, 1961.
- Electrophoresis buffer:
Prepare a stock solution 20X:
  Tris, 97 g
  Sodium acetate·$3H_2O$, 54.5 g
  Na_2EDTA, 7.4 g
Adjust pH to 7.8 with acetic acid and make up to 1 litre with distilled water.
- Polyacrylamide gel:
  Glass distilled water, 5.78 ml
  Electrophoresis buffer 3X, 4 ml
  Acrylamide 30 percent, bisacrylamide 0.8 percent, 2.4 ml
  Tetramethylethylenediamine (TEMED) 1 percent, 20 μl
  Ammonium persulphate (APS) 10 percent, 200 μl

PROCEDURE

1. Collect 10 to 15 g of cortical tissues from green or, preferably, mature canes and pulverize in a chilled mortar with liquid nitrogen.
2. Add 1 to 2 volumes of extraction buffer:
  2X STE, 45 ml

10 percent SDS, 15 ml

Phenol (water saturated), 25 ml

NH$_4$OH, 0.1 ml

Bentonite (40 mg per ml), 0.8 ml

2-Mercaptoethanol, 1 ml

3. Stir the mixture until amalgamation.

4. Add 25 ml of chloroform.

5. Stir at room temperature for 45 minutes.

6. Centrifuge at 8 000 rpm for 15 minutes.

7. Collect liquid phase by filtering through glass wool and make up to a volume of 100 ml.

8. Add 2 g of CF-11 cellulose powder.

9. Add 22 ml of 95 percent ethanol to reach a final alcohol concentration of approximately 17 percent. Stir for 45 minutes at room temperature.

10. Put the suspension in a glass column (alternatively a 10-ml plastic disposable syringe is suitable), allow the cellulose powder to set, then wash with not less than 100 ml STE containing 16.5 percent ethanol.

11. Drain cellulose by pumping air gently with a piston from the top of the column. Remove the piston, making sure that the cellulose pad does not break. Elute dsRNA by flushing the cellulose pad three times with 3 ml of STE each time. (To secure maximum recovery it is advisable to drain the pad each time by pumping air with a piston as above.)

12. Centrifuge the eluted liquid at 8 000 rpm for 3 minutes to eliminate residual cellulose.

13. Add 20 ml of 95 percent cold ethanol and 0.5 ml of 3 M sodium acetate pH 5 and stir thoroughly.

14. Incubate overnight at -20°C or for 3 hours at -70°C.

15. Centrifuge at 8 000 rpm for 30 minutes.

16. Save pellets and resuspend in 0.5 ml STE. Add 1 ml of cold 95 percent ethanol and 20 µl of 3 M sodium acetate pH 5, stir and leave standing at -70°C for 3 hours.

17. Centrifuge at 12 000 rpm for 15 minutes in an Eppendorf centrifuge, discard supernatant fluid and dry out pellets.

18. Resuspend pellets in 20 µl of electrophoretic buffer and add 5 µl of 0.3 percent bromophenol blue dissolved in 50 percent glycerol or sucrose solution.

19. Load on 6 percent polyacrylamide gel and apply a constant current of 30 mA for 6 hours or 20 mA overnight.

20. Remove gel from chamber and form, stain with ethidium bromide or silver nitrate and photograph.

REFERENCE

Fraenkel-Conrat, H., Singer, B. & Tsugita, A. 1961. Purification of viral RNA by means of bentonite. *Virology,* 14: 54-58.

ELECTROPHORESIS
Western blot
D. Boscia and G.P. Martelli

Western blotting is an electrophoretic procedure whereby the coat protein subunits of phloem-limited, non-mechanically transmissible grapevine viruses can be separated from other components of plant extracts and identified following the reaction with specific antisera and staining. When molecular weight markers are used, the molecular weight of protein subunits can be determined and compared with known values. Specific recognition relies upon the use of homologous immunoglobulins.

Operations and materials for electrophoresis are the same as described previously for detection of viroids.

MATERIALS AND SOLUTIONS
- Separating gel 12 percent:
    Acrylamide (30 percent) and bis-acrylamide (0.8 percent), 6 ml
    Tris buffer, 1.5 M, pH 8.8, 3.75 ml
    SDS 10 percent, 0.3 ml
    TEMED, 7.5 µl
    Ammonium persulphate (10 percent), 125 µl
    Distilled water, 4.8 ml
- Stacking gel 5 percent:
    Acrylamide (30 percent) and bis-acrylamide (0.8 percent), 1.67 ml
    Tris buffer, 0.4 M, pH 6.8, 1.23 ml
    SDS 10 percent, 0.1 ml
    TEMED, 5 µl
    Ammonium persulphate (10 percent), 150 µl
    Distilled water, 6.89 ml

- Degradation buffer 4X:
    Tris buffer, 0.4 M, pH 6.8, 100 ml
    SDS, 10 g
    2-Mercaptoethanol, 20 ml
    Sucrose, 20 g
    Bromophenol blue, 0.005 g
- Transfer buffer:
    Methanol 20 percent, 20 ml
    Tris, 25 mM, 0.3 g
    Glycine, 192 mM, 1.43 g
  Make up to 100 ml with distilled water and adjust pH to 8.3.
- PBS:
    NaCl, 8.0 g
    KH_2PO_4, 0.2 g
    NaH_2PO_4, 1.5 g
    KCl, 0.2 g
    Sodium azide, 0.2 g
  Dissolve in 1 000 ml of distilled water and adjust pH to 7.5.
- PBS-T:
  PBS plus 0.05 percent Tween 20
- TBS:
    Tris, 0.02 M, pH 7.5
    NaCl, 0.5 M
- TBS-T:
  TBS plus 0.05 percent Tween 20
- Citrate buffer 0.2 M:
    Citric acid, 13.5 g
    Sodium citrate, 11.0 g
  Dissolve in 1 000 ml of distilled water and adjust pH to 7.5.
- Enhancement solution:
    Dissolve 0.22 g hydroxiquinone in 22.5 ml

of citrate buffer. Immediately before use add 0.028 g of silver lactate dissolved in 2.5 ml of distilled water and mix thoroughly.
- Fixing solution:
Dilute fixing solution of Bio Rad kit 1:10 with distilled water.

PROCEDURE

1. Prepare polyacrylamide gels: 12 percent (separating gel) and 5 percent (stacking gel).
2. Put separating gel in the glass form, filling to about 2 cm from the top. Add distilled water gently and let stand for 30 minutes for polymerization.
3. Pour out distilled water, remove remaining water with filter paper, load stacking gel on top of polymerized separating gel and let stand for an additional 30 minutes for polymerization.
4. Connect power supply and make a blank electrophoretic run for 30 minutes at 30 V.
5. Prepare sample to be examined by grinding 2 to 5 g of cortical, petiolar or leaf vein tissues in a mortar with liquid nitrogen following the procedure outlined in the chapter on extraction of closteroviruses from grapevine tissues, below.
6. Resuspend high speed pellets in 200 ml of 4X degradation buffer and heat at 100°C for 2 minutes.
7. Place samples in the wells and apply 130 V constant current for about 1 hour until the stain bands reach the bottom of the gel.
8. Wearing rubber gloves, remove the gel and immerse it for 30 minutes in transfer buffer. At the same time, soak polyvinylidene difluoride membrane (PVDF) (Immobilon P-Millipore) in 100 percent methanol and the filter paper in distilled water, then move both to transfer buffer.
9. When gel is ready, gently remove Trans Blot cover and steel cathode (semi-dry transfer cell, Bio Rad)(Figures 310 and 311).

10. Place filter paper on to plate-shaped anode and roll a glass rod gently on top of it to remove air bubbles (Figure 312).
11. Place gel slab on filter paper, making sure that there are no air bubbles.
12. Place PVDF membrane on the gel slab, making sure that the gel is centred and that there are no air bubbles. Cover gel with another layer of filter paper and roll a glass rod gently over it to remove air bubbles (Figures 313 to 315).
13. Carefully replace steel cathode and Trans Blot cover, and connect the apparatus to power supply (Figure 316).
14. Apply current as follows (Figure 317):
- for minigels, 10 to 15 V for 10 to 15 minutes at an amperage not higher than 5.5 mA per cm^2 to avoid excess heating;
- for slab gels, 15 to 25 V for 30 minutes to 1 hour at an amperage of 3 mA per cm .
15. Remove PVDF membrane from Trans Blot and incubate for 1 hour at 37°C in PBS buffer containing 2 percent non-fat milk powder (2 g in 100 ml PBS buffer). Save 5 ml of this solution.
16. Place PVDF membrane in a polythene bag, and pipette into it 3 ml of PBS non-fat milk solution to which 3 µg per ml of immunoglobulins have been added.
17. Remove air bubbles from plastic bag and seal. Incubate at 37°C for 2 hours or overnight at 4°C.
18. Wash PVDF membrane three times for 10 minutes in PBS-T and twice for 10 minutes in TBS-T.
19. Incubate PVDF membrane with a suspension of protein A-gold conjugate (gold enhancement kit, Bio Rad) at room temperature in the dark for 1 to 24 hours. The length of incubation depends on the intensity of a signal constituted by a faint pink band. If no signal appears after 24 hours proceed to the following steps.
20. Wash PVDF membrane twice for 10 minutes in TBS-T, twice for 10 minutes in TBS, five

times for 1 minute in distilled water and once for 5 minutes in citrate buffer.

21. Prepare enhancement solution during the last washing.

22. Incubate PVDF membrane in enhancement solution at room temperature in the dark for 5 to 15 minutes. Rinse with distilled water.

23. Soak PVDF membrane in fixing solution for 5 minutes. Rinse with distilled water and dry with blotting paper.

24. Observe and record positions of dark-staining bands corresponding to dissociated viral coat protein subunits.

FIGURE 310
Trans Blot apparatus for electrophoretic
transfer of dissociated viral coat protein
subunits to PVDF membrane

FIGURE 313
Gel slab is placed on filter paper

FIGURE 311
Disassembled Trans Blot apparatus showing
plate-shaped electrodes

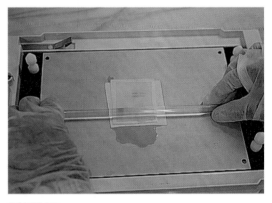

FIGURE 314
PVDF membrane is placed on gel slab and air bubbles
are removed by gentle rolling with a glass rod

FIGURE 312
Piece of wet filter paper placed on the anode. Plastic
containers on the left-hand side contain PVDF
membrane soaked in methanol and gel slab after
electrophoresis soaked in transfer buffer. The gel slab
has been stained on purpose to show protein band

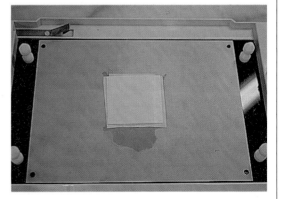

FIGURE 315
A piece of wet filter paper is placed on PVDF membrane;
blot is ready for electrophoretic transfer

FIGURE 316
Steel cathode is in place

FIGURE 317
Trans Blot apparatus connected to power
supply, ready for operation

Detection of viruses and viroids by molecular hybridization

G. Macquaire, T. Candresse and J. Dunez

PRINCIPLE

A viral particle is composed of nucleic acids [ribonucleic acid (RNA) or deoxyribonucleic acid (DNA)] and a capsid made up of several dozen to a thousand copies of coat protein subunit. In some cases, the virus possesses an envelope composed of viral proteins integrated in membranes deriving from the host cell. Serological techniques detect the virus by specific recognition of the coat protein by specific antibodies developed in animals against this protein. Molecular hybridization techniques detect viral nucleic acids by specific recognition of their nucleotide sequence.

Nucleic acids are long, linear polymers of nucleotide molecules. Each nucleotide is in turn composed of several elements: a nitrogen-containing base linked to a phosphate group and a sugar molecule (ribose for RNA and deoxyribose for DNA). DNA contains four different bases: adenine (A), guanine (G), cytosine (C) and thymine (T). In the case of RNA, thymine is replaced by uracil (U), the three other bases being the same.

DNA is usually found in a double-stranded configuration, i.e. two chains of DNA associate through specific base pairing (A pairs with T and C pairs with G). Base pairing is extremely specific and creates non-covalent hydrogen bonds that unite the molecules associated in this way. RNA is most commonly found in a single-stranded configuration but, like DNA, it possesses the capacity to form double-stranded structures through A-U and G-C pairing.

The specific pairing of the bases composing nucleic acids constitutes the basis for the formation of hybrids (double-stranded structure) between complementary molecules and thus for the use of molecular hybridization as a diagnostic technique.

Nucleic acid molecules differ from one another in the order and sequence of alignment of their nucleotides (= nucleotide sequence). Two molecules of complementary sequences will form double-stranded hybrids under suitable conditions. For example, TCGGCGTAT will pair with AGCCGCATA to make a DNA-DNA hybrid.

A probe used for virus detection in molecular hybridization experiments is a single-stranded nucleic acid molecule prepared from a viral nucleic acid, with a nucleotide sequence complementary to that of the target viral RNA molecule.

Thus a DNA probe with the sequence TCGGCGTAT will specifically detect RNA and DNA molecules with the respective sequences AGCCGCAUA and AGCCGCATA. An RNA probe with the same specificity would be UCGGCGUAU.

The molecular hybridization detection system presented here is based on a solid support hybridization, the samples being permanently immobilized on a nitrocellulose membrane

(Figure 318). We describe the technique using the two most frequently used types of probe:

- complementary DNA probes cloned in a plasmid vector;
- *in vitro* transcribed complementary RNA probes prepared from complementary DNA cloned into special purpose transcription plasmid vectors.

The probes can be labelled either radioactively or by incorporation of a non-radioactive marker such as biotin. The techniques for the determination of the probe-specific activity are described following the experimental protocol.

BUFFERS

See Table 9 for commercial souces of chemicals.

- Grinding buffer for viroids (GPS):
    - 200 mN glycine
    - 100 mM Na_2HPO_4
    - 600 mM NaCl
    - 1 percent SDS (sodium dodecyl sulphate)

    Adjust pH to 9.5, autoclave 20 minutes at 120°C, then add:
    - 0.1 percent DIECA (sodium diethyldithiocarbamate)
    - 0.1 mM DTT (dithiothreitol)
- Grinding buffer for viruses:
    - 50 mM sodium citrate, pH 8.3
    - 2 percent PVP (polyvinylpyrrolidone)

    Autoclave 20 minutes at 120°C, then add:
    - 1 mM EDTA
    - 20 mM DIECA
- Phenol/chloroform mixture:
    - 1 volume water-saturated phenol
    - 1 volume chloroform/pentanol (24 /1)
- 20X SSC buffer:
    - 3 M NaCl
    - 0.3 M sodium citrate

    Adjust pH to 7.0.
- Formaldehyde denaturation buffer:
    - 5X SSC

- 25 mN Na_2HPO_4
- 5X Denhart [0.1 percent each of bovine serum albumin (BSA), Ficoll and PVP 360]
- 50 percent deionized formamide
- 200 µg per ml of denatured calf thymus DNA
- DNA probes hybridization buffer:
    - 4 volumes DNA probes pre-hybridization buffer
    - 1 volume 50 percent dextran sulphate
- RNA probes pre-hybridization and hybridization buffer:
    - 50 percent formamide
    - 50 mM phosphate buffer pH 6.5
    - 5X SSC
    - 0.1 percent SDS
    - 1 mM EDTA
    - 0.05 percent Ficoll
    - 0.05 percent PVP 360
    - 200 µg per ml of denatured salmon sperm DNA
- Washing buffers:
    Washing buffer A:
    - 2X SSC
    - 0.1 percent SDS
    Washing buffer B:
    - 0.2X SSC
    - 0.1 percent SDS
    Washing buffer C:
    - 0.1X SSC
    - 0.1 percent SDS
- Development buffers for biotinylated probes:
    Buffer 1:
    - 100 mM Tris-HCl pH 7.5
    - 100 mM NaCl
    - 2 mM $MgCl_2$
    - 0.05 percent Triton X100
    Buffer 2:
    - Buffer 1 plus 3 percent BSA
    Buffer 3:
    - 100 mM Tris-HCl pH 9.5
    - 100 mM NaCl

TABLE 9
Commercial sources of chemicals

Chemical	Manufacturer[a]
BSA	Sigma No. A6793
PVP 360	Sigma No. P5288
Ficoll 400	Sigma No. F4377
Salmon sperm DNA	Sigma No. D1626
DTT	Sigma No. D9779
Dextran sulphate	Sigma No. D8906
SDS	Sigma No. L4390
Glycin	Sigma No. G4392
Deionized formamide	BRL No. 540-5515UB
EDTA (TITRIPLEX III)	Merck No. 8418
DIECA	Merck No. 6689
Phenol (analar grade)	Merck No. 10188
DNA detection system	BRL No. 530-8239SA
Bio-11 dUTP	BRL No. 520-9507SA
Nick translation kit	BRL No. 530-8160SB
Triton X100	Merck No. 11869
Chloroform	Merck No. 2442
Pentanol (isoamyl alcohol)	Merck No. 979
Calf thymus DNA	Sigma No. D8899

[a] Sigma Chemical Company Inc., PO Box 14508, St Louis, MO 63178, USA; BRL, Gibco BRL France, 14 rue des Osiers, BP 7050, 95051 Cergy Pontoise Cedex, France; Merck Schuchardt & Co., Eduard Buchner Strasse, D-8011 Hohenbrunn, Germany.

EXPERIMENTAL SET-UP

Schematically the technique can be divided into five steps:

- *Probe labelling*. This is achieved by incorporation of a labelled (radioactive or biotinylated) nucleotide triphosphate precursor during *in vitro* reactions (nick translation for DNA probes or transcription for RNA probes).
- *Sample preparation*. The nucleic acids present in the plant extract are immobilized on a nitrocellulose membrane by baking it for two hours at 80°C under vacuum.
- *Hybridization*. The labelled probe will form double-stranded structures under suitable conditions with the target nucleic acid immobilized on the membrane.
- *Washing(s)*. Non-hybridized probe molecules are removed by successive washings of the membrane under stringent conditions.
- *Hybrid detection*. For radioactive probes, this is achieved by contact of an X-ray film with the membrane (autoradiography), usually for 24 hours. In the case of biotinylated probes, three additional steps are required (Figure 319):
  1. incubation in the presence of streptavidin, which reacts specifically with the biotin molecules fixed on the probe;
  2. incubation in the presence of a biotinylated enzyme, which will be trapped by the streptavidin already retained on the membrane;

3. an enzymatic reaction that results in the formation of a coloured product at the fixing point of the complex probe biotin-streptavidin-biotinylated enzyme.

EXPERIMENTAL PROTOCOL
Probe labelling

DNA probes. Purified recombinant plasmid DNA is labelled (by incorporation of either ^{32}P sCTP, biotinylated dUTP or dCTP) by the technique of nick translation, using one of the several commercially available kits (e.g. BRL, Amersham).[1]

RNA probes. After linearization of the purified recombinant plasmid downstream of the viral cDNA with a suitable restriction endonuclease, labelled RNA is produced by *in vitro* transcription using one of several commercially available kits (e.g. Promega, Biotec, Boehringer).[2] Either ^{32}P or biotin-labelled CTP is usually incorporated.

Sample preparation

Many different plant samples can be used, consisting of leaves, stems, tubers, barks or fruits (Figure 320). There is no standard protocol; each protocol should be optimized for a given host/virus combination. We present here a technique for the detection of plum pox virus, with additional advice on detection of other pathogens when appropriate.

Sample grinding. One gram of plant sample is ground in 4 ml of grinding buffer using a pestle and mortar (or other apparatus such as an electric

press or Polytron homogenizer when available), (Figures 321 to 323). It is extremely important to use a buffer that will optimize the signal-to-noise ratio. The extract is then clarified by centrifugation for 10 minutes at 10 000 rpm (Figure 324). The samples can, if necessary, be deproteinized by including one volume of a 1:1 mixture of water-saturated phenol and chloroform during the grinding. This step is optional for the use of radioactive probes but necessary when using biotinylated probes.

Sample denaturation. The nucleic acids contained in the supernatant are then denatured if necessary, to ensure good binding of the nitrocellulose and availability of the sequences for hybridization (Figure 325). This step is important for the detection of viroids but of no utility for most viruses. In a small microcentrifuge tube, 50 µl of sample are added to 50 µl of formaldehyde denaturation buffer. The mixture is then incubated for 60 minutes at 60°C (the length of this incubation should be reduced for viruses). At this point, samples are ready for spotting on the membrane. They can also be stored for up to several months at -20°C. We have found that concentration of the nucleic acids present in the extract by ethanol precipitation is detrimental since it usually increases the non-specific background reactions; it is therefore not recommended.

Nitrocellulose membrane preparation. Soaking of the membrane in a high-salt solution is required for proper binding of nucleic acids in the samples. The membrane is first soaked for 2 minutes in pure distilled water and then equilibrated for 10 minutes in 20X SSC buffer (Figure 326).

Sample application and fixation. Next, 20 µl of sample are applied to the nitrocellulose membrane using a BRL "Hybri-dot" apparatus

---

[1] BRL, Gibco BRL France, 14 rue des Osiers, BP 7050, 95051 Cergy Pontoise Cedex, France; Amersham Corp., 2636 S. Clearbrook Dr., Arlington Heights, IL, USA.
[2] Promega Corp., 2800 S. Fish Hatchery Road, Madison, WI 53711-5305, USA; Biotechnie France S.A., Soaris 139, 94524 Rungis Cedex, France; Boehringer Mannheim GmbH, Pasettistrasse 64, A-1201, Vienna, Austria.

(Figures 327 to 330). Alternatively, 3 to 5 μl of sample can be applied directly (using a micropipette) to nitrocellulose that has been air-dried after soaking in 20X SSC. The membrane is then dried at room temperature (Figure 331) and baked for a further 2 hours at 80°C under vacuum to ensure stable binding of the nucleic acids to the nitrocellulose membrane (Figures 332 and 333). This can conveniently be achieved by using an electrophoresis slab gel drier or a vacuum oven. At this point, the membranes can be directly processed or sealed in a plastic bag (Figures 334 and 335) and stored (at 4°C or -20°C) for up to several months.

Hybridization reaction

Pre-hybridization. In order to prevent non-specific binding of the probe to the membranes, they are pre-incubated in the hybridization mixture (pre-hybridization). The membranes are sealed in a plastic bag in the presence of 1 ml of hybridization buffer for each 10 cm^2 of membrane, taking care to avoid trapping any air bubbles (Figure 336). The bag is then incubated for 2 to 4 hours in a water bath at 42°C (Figure 337).

Probe denaturation. This step is included to remove any secondary structure of the probe and is especially important for DNA probes which are essentially double-stranded after the labelling reaction. A suitable quantity of probe is placed in a small disposable tube and incubated for 10 minutes (DNA probe) or 3 minutes (RNA probe) at 100°C in a bath of boiling water (Figure 338) and then quickly chilled by placing the tube in an ice-bucket.

Hybridization. The pre-hybridization buffer is discarded and replaced by the hybridization buffer to which the denatured probe has been added. Use approximately 1 ml of buffer containing

radioactive probe of 1 to 2 x 10^6 cpm per ml or 200 ng per ml of biotinylated probe per 15 cm^2 of membrane (See techniques for determination of probe-specific activity, below). The plastic bag is then resealed and incubated in a water-bath overnight at 50°C (Figures 339 and 340).

Washing

After hybridization is completed, the membrane is removed from the plastic bag and washed in a small plastic tray (Figure 341). After washing, the nitrocellulose membranes should be air-dried at room temperature (Figure 342).

DNA probes. Wash at room temperature for 5 minutes in three changes of washing buffer A, then proceed with two 15-minutes washes at 50°C in washing buffer B.

RNA probes. Carry out four 20-minute washes at 60°C in washing buffer C.

Hybrid detection

Radioactive probes. An X-ray film (Kodak XAR or equivalent) is exposed to the membrane for 24 hours at -70°C using intensifying screens (Figures 343 and 344). After autoradiography, the film is developed using Kodak LX 24 developer and Ilford Hypam fixer (Figure 345). Within the limits of linearity of the response of the film, the intensity of the spots is proportional to the concentration of viral RNA present on the membrane. No non-specific signal should be obtained with healthy plant controls.

Biotinylated probes. Several commercially available kits can be used for the detection of biotinylated probes (e.g. BRL). The composition of the buffers is given above.

The membranes are first soaked for 1 minute at room temperature in buffer 1, then for 20 minutes at 42°C in buffer 2 to saturate the protein-fixing sites on the membrane. They are then dried and baked for 10 to 20 minutes at 80°C under vacuum.

Following the treatment, the membranes are rehydrated for 10 minutes in buffer 2 and then incubated on a Petri dish in the streptavidin solution: 6 µl of a 1-mg per ml streptavidin solution diluted in 3 ml of buffer 1. Incubate for 10 minutes at room temperature, shaking occasionally.

The membranes are then washed well with at least three changes of buffer for 3 minutes each time. Incubate on a Petri dish with 3 ml of buffer 1 containing 3 µl of a solution of bio-tinylated polymers of alkaline phosphatase (polyAP) at 1 mg per ml. Incubate for 10 minutes at room temperature with occasional shaking.

Wash abundantly with two changes of buffer 1 and then with two changes of buffer 3. The developing solution should be prepared at the last moment in the following way: add 33 µl of the nitro-blue tetrazolium solution to 7.5 ml of buffer 3. Mix thoroughly, then add 25 µl of the 5-bromo-4-chloro-3-indolyl phosphate (BCIP) solution mix. Incubate the membrane in this solution in a sealed plastic bag protected from light.

Maximum colour development is usually achieved within 4 hours. To stop the development, simply wash the membrane in 20 mM Tris-HCl pH 7.5, 5 mM EDTA. The dried membranes can then be stored for several months in the dark to preserve the colour.

DETERMINATION OF THE PROBE-SPECIFIC ACTIVITY

Following the labelling reaction, the radioactive DNA probe (1 µg) is precipitated with ethanol, freed from the unincorporated labelled nucleotides by several 70 percent ethanol washes, dried and finally taken up in 100 µl of sterilized distilled water. Then 2 µl of the probe are mixed with 3 ml of a 10 percent trichloroacetic acid (TCA) solution along with 10 µl of 3 mg per ml calf thymus DNA used as a carrier. The mixture is left for 30 minutes at 0°C and then filtered through a Whatman GF/C fibreglass filter. The filter is rinsed with 20 ml of a 5 percent TCA solution and then with 5 ml of ethanol before being dried. The radioactivity retained on the filter is then determined by liquid scintillation counting. The specific activity, in cpm per µg, is given by cpm $\times$ 100/2.

Besides determining the specific activity of the probe, this technique can also help to calculate how much of the probe should be added to the hybridization reaction. Radioactive RNA probes can be counted in the same way.

For biotinylated probes, the result of the labelling reaction can be estimated by spotting dilutions of the probe on a membrane and comparing with a standard of known activity provided in the labelling kit.

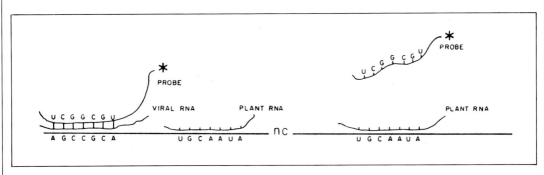

FIGURE 318
Schematic representation of the hybridization of a probe to nucleic acids immobilized on a nitrocellulose membrane
(left) Infected sample containing normal plant cell RNAs and viral RNA (the target sequence to which the RNA probe hybridizes). Because of this hybridization the labelled probe will be retained on the membrane
(right) Healthy sample containing only normal plant cell RNAs. The probe cannot hybridize with any sequence; it will not be retained on the membrane and will be eliminated upon washing

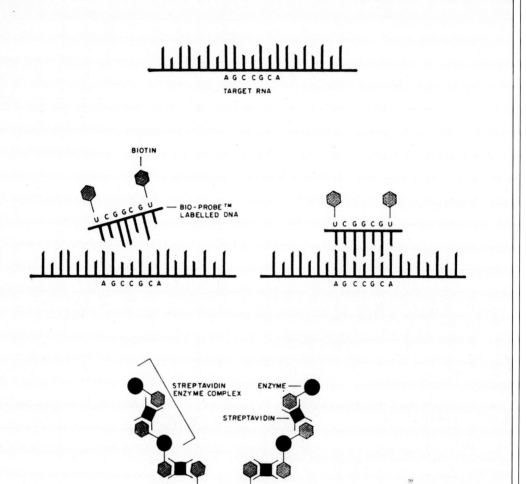

FIGURE 319
Schematic representation of the system used for the detection of biotinylated probes

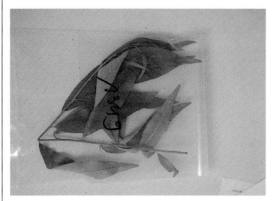

FIGURE 320
Plant sample consisting of leaves, bark, roots, tubers, fruits etc.

FIGURE 321
Plant sap is extracted using an electric press

FIGURE 322
A drop of sap is added to a grinding buffer contained in the microcentrifuge tube

FIGURE 323
Sap and grinding buffer are mixed using a vortex

FIGURE 324
Low-speed centrifugation (10 000 rpm for 10 minutes) is carried out to pellet the plant cell debris and to separate the phases if a phenol deproteinization step is included

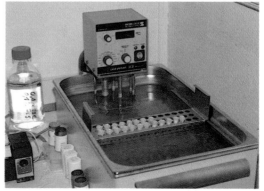

FIGURE 325
For viroids, the supernatants are supplemented with formaldehyde and incubated for 60 minutes at 60°C (omit this step for most viruses)

FIGURE 326
In the meantime, one nitrocellulose membrane and three
sheets of Whatman 3MM paper are cut, soaked in water and
equilibrated in 20X SCC buffer

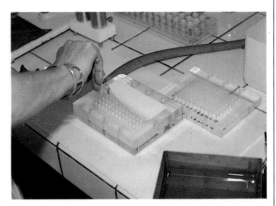

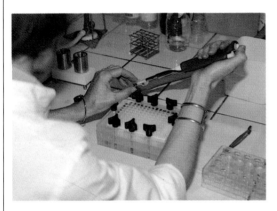

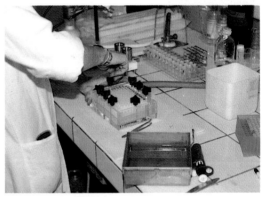

FIGURES 327 and 328
One 3MM filter and the nitrocellulose are placed in the BRL
Hybri-dot system (or equivalent)

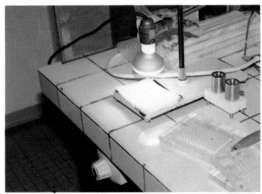

FIGURE 331
The membrane is taken out of the blotting apparatus and
air-dried

FIGURES 329 and 330
Next, 20 μl of extract are spotted on the membrane while
gentle vacuum is applied

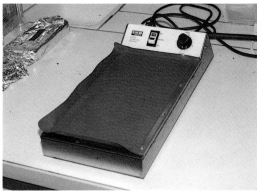

FIGURES 332 and 333
The membrane is sandwiched between the two remaining filters and baked in vacuum for 2 hours in a gel dryer

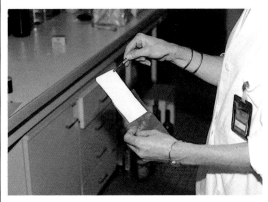
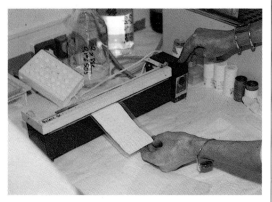

FIGURES 334 and 335
After drying, the membrane is placed in a plastic bag which is sealed on three sides

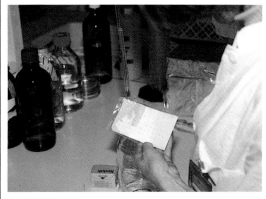

FIGURE 336
The pre-hybridization buffer is added to the plastic bag and the bag is completely sealed

FIGURE 337
Pre-hybridization is carried out by incubating the bags for 2 to 4 hours in a water bath under suitable conditions

FIGURE 338
The probe is denatured for a few minutes in a bath of boiling water

FIGURE 339
The bag is cut open and the pre-hybridization buffer is discarded and replaced by the hybridization buffer containing the denatured probe. The bag is then resealed

FIGURE 340
Hybridization is carried out by incubating the bag overnight at the desired temperature in a water-bath

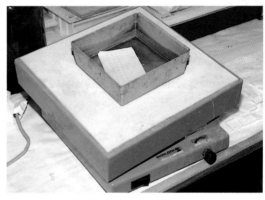

FIGURE 341
After hybridization, the membrane is removed from the bag and washed in several changes of washing buffer

FIGURE 342
After washing, the membrane is air-dried

FIGURES 343 and 344
An X-ray film is exposed to the membrane

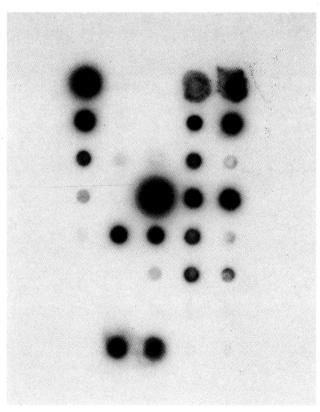

FIGURE 345
After autoradiography is completed, the X-ray film is developed and
fixed

OTHER METHODS

Extraction of closteroviruses from grapevine tissues

V. Savino

Closteroviruses are the cause of, or are associated with, diseases of grapevines such as leafroll and rugose wood. Except for grapevine clostero-viruses A and B (GVA and GVB), which can occasionally be transmitted by inoculation of sap to herbaceous hosts, none of the other clostero-viruses described to date are mechanically transmissible. Their isolation relies on extraction from naturally infected grapevine tissues.

PROCEDURE

The following procedure gives reproducible results.

1. Collect samples (10 g each) consisting of adult symptomatic leaves or cortex scrapings from mature dormant canes. Leaves may be placed in a freezer (-20°C) for 24 hours or longer before processing. Cortex shavings are recommended for American rootstocks, especially for pure *Vitis rupestris* and its hybrids, because extractions from leaves are negative.

2. Grind with pestle and mortar in liquid nitrogen until tissues are pulverized.

3. Add 5 volumes of the following extraction buffer:

> 0.5 M Tris-HCl, pH 8.2
> 4 percent Polyclar AT
> 0.5 percent bentonite
> 1 percent Triton X-100
> 0.2 percent 2-mercaptoethanol

It is advisable to prepare fresh buffer each time and stir it for 30 minutes before use.

4. Stir in the cold (4°C) for 15 minutes.

5. Centrifuge at 5 000 rpm for 20 minutes.

6. Discard pellets. Save and filter supernatant liquid.

7. Centrifuge at 27 000 rpm for 2 hours.

8. Discard supernatant liquid. Resuspend pellets in a few drops of 0.02 M Tris-HCl buffer, pH 8.2, containing 0.01 M magnesium chloride.

9. Place a drop of the suspension on a dental wax bar or hydrophobic filter paper.

10. Float a freshly carbon-coated electron microscope grid on the drop for 15 minutes.

11. Stain with uranyl acetate and observe under the electron microscope.

OTHER METHODS

Extraction of phloem-limited isometric viruses from grapevine tissues

G.P. Martelli

Phloem-limited viruses with isometric particles 25 to 30 nm in diameter are the cause of or are associated with diseases of the grapevine such as fleck, ajinashika disease and grapevine stunt. The latter two diseases have been reported only from Japan. None of these viruses are transmissible by inoculation of sap to herbaceous hosts. Their isolation relies upon extraction from naturally infected grapevine tissues.

PROCEDURE

A procedure that yields reproducible results for the recovery of the virus associated with fleck is the following:

1. Collect samples (10 g each) consisting of main veins and petioles or young succulent roots.
2. Grind with quartz sand in a mortar in the presence of 1 or 2 volumes of 0.05 M phosphate buffer pH 7.2 containing 5 mM 2-mercaptoethanol, 5 mM DIECA, 5 mM EDTA and 5 g per litre polyethylene glycol 6 000 (PEG).
3. Add 4 volumes of buffer to homogenate plus pectinase (10 g per litre) and cellulase (20 g per litre) and incubate at room temperature in the dark for 12 to 14 hours.
4. Strain through cheesecloth and centrifuge at 6 000g for 20 minutes.
5. Collect supernatant fluid, add 10 percent PEG and 1 percent NaCl while stirring, and incubate for 2 hours at 4°C.
6. Centrifuge at 10 000g for 20 minutes.
7. Discard supernatant and resuspend pellets in 0.05 M phosphate buffer, pH 6.8, containing 5 mM 2-mercaptoethanol, 5 mM EDTA and 0.2 M NaCl.
8. Stain with uranyl acetate and observe under the electron microscope.

OTHER METHODS

Isolation and culture of *Xylella fastidiosa*

D.A. Golino

PROCEDURE

1. Prepare culture medium (PD3 according to Davis, Whitcomb and Gillaspie, 1981) as follows (quantities per litre):

Pancreatic digest of casein (tryptone, Difco), 4 g

Papaic digest of soy meal (Soytone, Difco or Phytone peptone, BBL), 2 g

Trisodium citrate, 1 g

Disodium succinate, 1 g

Hemin chloride stock (0.1 percent hemin chloride dissolved in 0.05 N NaOH), 10 ml

$MgSO_4 \cdot 7H_2O$, 1 g

K_2HPO_4, 1.5 g

KH_2PO_4, 1 g

Soluble potato starch, 2 g

Bacteriological agar (Difco), 15 g

Dissolve, autoclave and pour an appropriate amount into sterile Petri dishes.

2. Collect petioles from symptomatic grapevine leaves.

3. Cut petioles into fragments 2 to 3 cm long.

4. Surface sterilize by immersing for 5 to 10 minutes in 0.5 percent water solution of sodium hypochloride.

5. Rinse in three changes of sterilized distilled water under a sterile hood.

6. Express sap from petiole fragments by squeezing with sterile pliers or forceps and blot sap droplets directly on to culture medium. Alternatively, chop tissues finely with a razorblade in a sterile Petri dish or grind tissues in a sterile mortar in the presence of 1.2 ml of sterile distilled water. Streak the slurry on to culture medium with a loop.

7. Incubate at 28°C under aerobic conditions.

REFERENCE

Davis, M.J., Whitcomb, R.F. & Gillaspie, A.G.M. 1981. Fastidious bacteria of plants and insects (including so-called rickettsia-like bacteria). *In* M.P. Starr, H.O. Stolp, H.G. Truper, A. Balows & H.G. Schlegel, eds. *The prokaryotes: a handbook on habitats, isolation and identification of bacteria*, p. 2171-2188. Berlin, Springer-Verlag.

Part IV

Appendixes

Laboratory equipment needed for selected diagnostic procedures

R.F. Lee

The importance of the plant laboratory cannot be overemphasized. The ability to grow indicator plants that can be effectively used to detect the presence of viruses is fundamental to the success of a clean stock programme. The availability of a laboratory and basic laboratory equipment for certain diagnostic procedures is also important. When a virus-detection laboratory is being established, the question soon arises as to what equipment is needed, where it can be bought and the cost. In this section the basic laboratory equipment and the specialized equipment needed for three commonly used diagnostic procedures (ELISA, electrophoresis and culturing) are given, with the recent price range in the United States.[1]

Much of the following equipment may be obtained from local suppliers. As trade names may vary, the basic requirements are given rather than specific brands and model numbers. Each year the worldwide *Laboratory buyer's guide* is published for purchasers of laboratory equipment, chemicals and reagents. It includes product listings, a manufacturers' directory and a list of laboratory dealers around the world. The *Laboratory buyer's guide* may be obtained by sending US$25 to International Scientific Communications Inc., 30 Controls Dr., PO Box 870, Shelton, CT 06484-0870, USA [Telephone (203)926-9300, Telex 964292, Fax (203)926-9310].

While more specialized equipment is needed for some of the diagnostic procedures, there are a few basic requirements for any laboratory.

Size

A minimum area of about 55 m^2 is needed for a diagnostic laboratory. Thought needs to be given to designing effective work areas so that equipment and items needed for a particular procedure are conveniently located. With a proper arrangement of work areas, up to four persons can work effectively in a laboratory of this size. There should be a fume hood available, plus a sink for dishwashing and cabinets for storage of glassware, equipment and reagents.

Electrical power

A source of stable electrical power is necessary. If the power source is subject to interruptions or may be down for long periods, a generator should be available to power important equipment and to permit work to continue. Most electrical equipment manufactured in the United States for domestic use is 60 Hz (cycles per second). It is important that power sources and equipment be able to handle the cycle sequence in the country of use.

Air conditioning

Electronic equipment and instruments depending upon optical filters and diffraction gratings for

---

[1] Specific and more detailed information on equipment requirements are given under many of the procedures in Part III of this handbook.

operation, such as spectrophotometers and ELISA plate readers, are sensitive to high temperatures and high humidity. Fungi can ruin filters and gratings essential to the operation of spectrophotometers, and high humidity will cause corrosion of even solid-state electronic circuitry.

Deionized water supply

Costs for this vary depending on local water quality; usually a mixed bed resin filter with a resistance meter is adequate. Culturing of micro-organisms requires very high-quality water and a still may be needed.

Glassware

Although a single piece of glassware represents a small part of the total cost of equipping a laboratory, sufficient glassware is fundamental for effective operation of a diagnostic laboratory. A complete set of volumetric flasks and graduated cylinders needs to be available, as well as flasks, beakers, Petri dishes, test-tubes, centrifuge tubes, test-tube racks, ice containers, spatulas and containers for buffers and solutions. If 500 samples are to be run for ELISA, there need to be 500 tubes with racks to hold the samples. Allow US$2 000 to $3 000 for glassware to equip a diagnostic laboratory.

pH meter

The laboratory must have a pH meter capable of reading to 0.1 pH unit, with reference standards so that the instrument can be calibrated and tested. Select an electrode that will measure pH of Tris buffers. Prices start at about $100 for a hand-held unit and rise depending on the features selected.

Refrigerator

A refrigerator is needed for storage of reagents, chemicals and seed at 4°C. It should be large enough to hold racks of ELISA tubes until the ELISA test is complete. A refrigerator with sliding glass doors is desirable. If large enough, it can be used to run native gels for PAGE with the power supply located outside. Similarly, small centrifuges can be run inside a larger refrigerator and observed through the glass doors.

Freezer

A freezer is needed for storage of reagents, chemicals and samples at -2°C and to produce ice needed for use in ice-baths. Select a freezer that does not have an automatic defrost cycle; the heating cycles of an automatic defrost unit will cause a more rapid breakdown of reagents and nucleic acid preparations stored in the unit.

Balance

A balance is needed with sensitivity of at least 0.01 g. A top-loading electronic balance with dual-range sensitivity of 0.01 to 120 g and 0.1 to 1 200 g is ideal. Prices start at about $1 400.

Means of sterilizing equipment

An autoclave is useful for sterilizing equipment, glassware and reagents used for many diagnostic procedures, and it is absolutely essential for culturing. A large pressure-cooker will serve this purpose and is relatively inexpensive, while a small automated autoclave (prices begin at about $3 000) is convenient if much culturing is to be done. For sterilization of glassware, a glassware oven will be suitable.

Centrifuge

Ideally the centrifuge should be a refrigerated model with a rotor capable of holding 50-ml centrifuge tubes and adapters to accommodate smaller tubes. Prices of such centrifuges start at about $10 000. At the simplest, a clinical centrifuge with interchangeable rotors to accommodate different sizes of centrifuge tubes will suffice in many procedures (at a cost beginning at about $1 000). If the centrifuge is small enough, it can be operated in the bottom of a refrigerator for cooling. Often changes must be made to established protocols because of limitations in the ability to carry out the centrifugation steps. In some instances the sample must be divided into two tubes because the total volume specified will not fit into one tube, or the centrifuge time must be increased to account for a centrifugal force lower than that specified in the protocol.

Magnetic stirrer and hot plate

These can be purchased as separate units at costs starting at about $125 each or as a combined unit at costs starting at about $300.

SPECIALIZED EQUIPMENT FOR ELISA

Antisera and/or conjugates needed for ELISA are commonly available from type-culture collections, commercial sources or fellow research scientists. A low-speed centrifuge, UV-visible spectrophotometer and simple chromatography are needed if IgG and conjugates are to be prepared. Other equipment needed for ELISA is the following.

Repeating pipette

A repeating pipette which allows multiple pipettings with good accuracy is essential. At the minimum, a fixed volume (200 µl) pipette is required (cost about $125 to $ 250). It is highly desirable to have a set of three adjustable pipettes (0 to 20 µl, 20 to 200 µl and 200 to 1 000 µl) at a cost of about $125 to $250 each. If much ELISA is to be performed, a multichannel adjustable (50 to 250 µl) pipette should be considered at costs beginning at about $500. Microcapillary tubes or Drummond pipettes can be used to measure small volumes of IgG and conjugate if the 0 to 20 µl pipette is not available.

Grinding equipment

Although samples may be homogenized with pestle and mortar, a mechanical device is desirable if large numbers of samples are to be assayed. Dispersion homogenizers with a generator shaft of 15 to 25 mm diameter are commonly used, e.g. Polytron, Tissumizer, VirTis or Tissu-Tearor in a price range of $700 to $2 500. Rollers and stomacher devices work for some applications.

Evaluation of results

ELISA results can be estimated visually, but it is difficult to determine weak reactions, especially when there is a background. Photometric measurements must be made to obtain quantitative data. Aliquots of the reaction can be diluted in water and the adsorbance read in a regular spectrophotometer. Manually operated ELISA plate readers, available from about $4 000 upward, can quickly and accurately read an ELISA plate. If ELISA is to be performed on a sizeable scale, consideration should be given to purchasing an automated ELISA plate reader with an RS 232 port linked to a personal computer. Costs for such a system begin at about $20 000.

SPECIALIZED EQUIPMENT FOR ELECTROPHORESIS
Electrophoresis apparatus

Electrophoresis is often used to diagnose viroids or for the analysis of dsRNAs to detect the presence of viruses. Electrophoresis for these procedures is most commonly performed on vertical slab gels, although tube gel electrophoresis can also be used. Many models of both vertical slab and tube gel apparatus are sold commercially, with prices starting from about $225 for a small apparatus to $1 500 for larger units.

When the apparatus is being selected, its potential use should be taken into account. The same electrophoresis apparatus can also be used for SDS polyacrylamide gel electrophoresis for protein analyses and for non-denaturing gels for isozyme analysis. The larger units will cost more initially, will need more reagents because of the larger volume and will require longer run times, but they offer better resolution. The small units offer speed and lower cost, but often lack a high degree of resolution. For most diagnostic applications, the small unit is satisfactory. An alternative to purchasing a commercial unit is to custom-make a unit from Perspex. The most expensive component is the platinum wire needed for the electrode. This can be bought from an electron microscopy supply catalogue or an electronics shop.

Power supply

A stable power supply is essential for electrophoresis. For versatility, the power supply should be capable of running at constant voltage (0 to 500 V range or greater) or at constant current (0 to 400 mA range or greater). Cost of equipment that provides this range of control over the power supply begins at about $900 and increases depending on the model and features desired. Care must be taken to match the frequency (Hz) of equipment to that of the main electricity supply.

Vacuum pump and dessicator

These are often used to de-gas acrylamide solutions before pouring gels. In addition, nucleic acid preparations are usually dried in a vacuum after being collected as ethanal-precipitated pellets before being used for electrophoresis. A dessicator costs from about $50 depending on size. Hand vacuum pumps are available at costs beginning at about $30; water aspirators can also be used. An electric vacuum pump, which offers more versatility, costs from about $250 upward.

Visualization of samples on gels

The usual method used to detect nucleic acids (such as viroids and/or dsRNAs) on gels after electrophoresis is by staining with ethidium bromide, then viewing over an ultraviolet (UV) transilluminator. The nucleic acids fluoresce, and the resultant diagnostic bands on the gel can be visualized and photographed. Visualization by this method allows subsequent manipulations of the nucleic acids, such as infectivity assays, electrophoresis on denaturing gels and preparation of probes (Part III). A UV transilluminator with 302 nm wavelength is recommended for use with ethidium bromide staining, as hand-held short-wave UV lights do not have enough light intensity to visualize any but the strongest of gel bands. Prices for a UV transilluminator start at about $1 100 and increase in price as the filter size increases.

An alternative to viewing nucleic acids by fluorescence over UV light is to silver-stain the gels (Part III). Silver-staining is as sensitive as or more sensitive than ethidium bromide staining and eliminates the need for a UV transilluminator. However, silver-staining inactivates the nucleic acid and immobilizes it in the gel, which does not permit subsequent manipulation of the nucleic acid.

Documentation of the gel requires a means of photographing the gel. A UV-1 filter between the camera and the UV light source is commonly used. Polaroid cameras are routinely used. Instant cameras with a hood that fits over the gel on the UV transilluminator and a fixed focal length are now available with prices beginning at about $400. More elaborate Polaroid set-ups begin at about $4 000. To minimize investment and film costs, an SLR 35-mm camera with macro-focus lens and auto exposure can be used with black-and-white film such as Kodak Contrast Process Pan or equivalent. After the gel has been photographed, the film can be processed immediately before moving the gel (usually 15 to 20 minutes of darkroom time) to verify that the gel bands have photographed well and that the focus is satisfactory.

SPECIALIZED EQUIPMENT FOR CULTURING

Culturing is often used to verify the presence of citrus stubborn and other disorders caused by harmful prokaryotes such as citrus greening and Pierce's disease of grapevine.

Autoclave

An autoclave is essential for culture work. This can be as simple as a large pressure-cooker, but if a lot of culture work is to be done the additional cost of an automated autoclave may be justified.

Transfer hood

It is desirable to have a transfer hood equipped with filtered air and a flame to sterilize transfer loops. In more arid climates, an open work space in a room with no air movement can be lined with wet paper towels and the surface sterilized with 70 percent alcohol for culture work. More contaminations will occur than when a hood is used, however, and this method will not be satisfactory at all in hot, humid climates.

Incubators

These need to be available. Shaking capability can easily be obtained by placing small shakers in incubator cabinets. Costs of small shakers start at about $300.

Microscopes

A light microscope with phase contrast and good optics is needed to verify the presence of spiroplasmas and harmful prokaryotes from culture. Costs for a suitable microscope begin at about $2 000 and increase as quality increases. A stereoscope is needed as well, as an aid to culture work and also for general use and shoot-tip grafting. Costs for a stereoscope begin at about $1 200.

Glossary

Acquisition feeding

The feeding period during which an insect ingests sap containing the pathogen

Alkaline phosphatase

An enzyme that hydrolyses certain phosphate-containing compounds under alkaline conditions; commonly obtained from calf intestine mucosa

Antibody

A protein formed in blood serum in response to stimulation by an antigen. Antibodies are specific for their respective antigens, and antigens and antibodies are mutually attracted

Antibody-antigen complex

The reaction or attraction formed when reaction antigens meet their corresponding antibodies or vice versa. This strong attraction is the basis of all immunodetection systems

Antigen

A substance, often a virus or bacterium, that stimulates production of antibodies in an animal. Specifically, the corresponding molecule to the antibody in a serological test

Autoradiography

The technique or process of making a picture revealing the presence of radioactive material,

the film being laid directly on the object to be tested. Frequently used for detection of radioactivity following hybridization, by exposing filter paper to sensitive X-ray film

Biotin

A water-soluble vitamin of the B complex widely distributed in plant and animal tissue; it binds strongly to a glycoprotein named avidin.

Biotin derivatives of deoxyribonucleotides are incorporated into probe DNA by nick translation (see Part III). After hybridization the biotin can then be detected using streptavidin-fluorescein complexes. The streptavidin binds to the biotin by one of the strongest known biological interactions. The enzyme (usually peroxidase or alkaline phosphatase) is then reacted with its substrate which gives a coloured product; fluorescein is detected by fluorescence under light of a certain wavelength

Biotinylated compound

A compound to which the small vitamin biotin has been attached. Antibodies, enzymes and nucleic acids can be labelled with biotin

Biotinylated enzyme

An enzyme coupled chemically to biotin

Biotinylated probe

A DNA probe in which certain bases have been modified by chemical coupling of biotin

Blot

Verb: to transfer DNA, RNA or protein to an immobilizing matrix such as DMB-paper, nitrocellulose or nylon membranes

---

Note: Glossary prepared by C.N. Roistacher and G.P. Martelli. Some of the technical terms defined are from Oliver, S.G. & Ward, J.M., *A dictionary of genetic engineering*, Cambridge University Press. Appreciation is also expressed to M. Bar-Joseph, J.A. Dodds, S.M. Garnsey, J.V. Leary and J.S. Semancik for their contributions.

Noun: the autoradiograph produced during the Southern or Northern blotting procedure

BRL Hybri-dot

A commercially available kit for applying small volumes of extracts to a membrane to test for the presence of viruses or viroids by hybridization

Bud-graft inoculation

A bud, blind bud or chip bud cut from a stem of the plant or tree to be indexed and grafted to an indicator plant or tree

Bud-union crease

A line, ridge or fold, usually discoloured as brown, yellowish-brown or reddish-brown, and formed at the bud union. It is readily observed when the outer bark is removed. Some bud-union creases are caused by pathogens and others by incompatibility of rootstocks and scions

cDNA

Complementary DNA. The DNA complement of an RNA sequence. It is synthesized by the enzyme RNA-primed DNA polymerase or reverse transcriptase. The single-stranded DNA product of this enzyme (the reverse transcript) may be converted into the double-stranded form by DNA-primed DNA polymerase and inserted into a suitable vector to make a cDNA clone. cDNA cloning is commonly used to achieve the expression of mammalian genes in bacteria or yeast

cDNA probe

A radioactive specific DNA sequence used to detect complementary sequences of RNA or DNA (see cDNA and Probe)

Certification programme

A programme developed by a country, state, university or research centre for ensuring that selected budwood distributed to the growers is free of graft-transmissible pathogens and that the fruit is true-to-type. These pathogen-free certified trees are usually registered, and budwood issued from these mother or foundation block trees can be used to produce additional buds in an increase block for the development of certified trees

CF-11

A fibrous, graded cellulose powder sold by the Whatman Company

Chip bud

A piece of bark tissue used for graft inoculation. It is used when the bark of receptor indicator plants does not slip or open up to accept a bud or blind bud

Chromatography

The separation of mixtures of chemicals, compounds, proteins, macromolecules, etc. into their constituents or components by preferential adsorption by a solid such as a column of cellulose or by filter paper or gel

Clone

A budline derived from a single parent source by propagation from that source

Complementary

See cDNA. A nucleic acid sequence is said to be complementary to another if it is able to form a perfect hydrogen-bonded duplex with it, according to the Watson-Crick rules of base pairing. A viral genomic ssRNA is complementary to negative-sense ssRNA from which it is transcribed

Conjugated molecule

As used in ELISA, antibody and enzyme proteins combined to form an enzyme-labelled antibody.

The enzyme can then be detected colorimetrically, and the colour produced will fairly precisely indicate the amount of virus present

Corky bark

Abnormal condition of grapevine bark due to excessive cork production, which confers upon the trunk a spongy texture and a rough appearance, as with corky bark disease (see Rugose wood)

Coulure

Grapevine disorder whereby berries fail to set, yielding straggly clusters. It can be induced and/or enhanced by viral infections (nepoviruses in particular)

DEAE cellulose column

A plastic or glass tube open at the top and fitted with a stopcock on the bottom, containing diethylaminoethyl cellulose

Denature

To so modify (a protein) by heat, acid or alkali that it retains its primary structure but no longer has all its original properties

Deproteination

To remove and separate proteins from other macromolecules from samples to be tested by hybridization. This is normally achieved by phenol extraction or by treating with a protein-digesting enzyme

Dialysis

A procedure using a membrane to separate various components in solution in accordance with their ability to pass through the membrane

Dicing

Cutting of tissue into small segments using a sharp knife or razor-blade. In the ELISA technique, leaf or bark segments are diced or cut up prior to grinding

DNA

Deoxyribonucleic acid. Any of a class of nucleic acids that contain deoxyribose, found chiefly in the nucleus of cells. Functions in the transference of genetic characteristics and in the synthesis of protein

DNA probe

A probe for detection of specific nucleic acid segments (see Probe)

Dot-blot

A procedure used to determine the presence and concentration of a particular RNA or DNA species. Different concentrations of the non-radioactive nucleic acids are denatured and applied as a dot to nitrocellulose paper or other support matrix. This is then hybridized with the radioactive complementary probe under study. After autoradiography, the intensities of the radioactive images formed are quantified and compared to a control series to determine the concentration of the non-radioactive molecule

dPAGE

Denaturing polyacrylamide gel electrophoresis. Electrophoresis in a gel formed from polyacrylamide in the presence of a chemical agent such as urea (8 M) or heat; the technique minimizes the effects of the secondary and tertiary structure of the molecule on electrophoretic mobility

Electro-blot membrane

A solid charged medium on to which a molecule is fixed as a result of electrophoresis from a source medium. This can be nitrocellulose or a charged nylon matrix

Electro-blotting

The electrophoretic transfer of macromolecules (DNA, RNA or protein) from a gel in which they have been separated to a support matrix such as a nitrocellulose or charged nylon sheet. An alternative to the capillary transfer usually used in techniques such as Southern and Northern blotting

Electro-elution

Removal of adsorbed material from an adsorbent by use of an electric field; or recovery of a charged molecular species by electrophoretic migration from a source medium such as a polyacrylamide gel to a liquid medium in which concentration of the species can be accomplished

Electrophoretic techniques

Techniques of separating components suspended in a fluid media or gel by the influence of an electric field

Electro-transfer

The movement of a charged molecule from one medium to a second by migration in an electric field

ELISA

Enzyme-linked immunosorbent assay. Two antibody preparations are commonly used in ELISA. The primary antibody binds the antigen, which is itself bound by the second antibody. The second antibody is linked to any enzyme whose activity is easily monitored, i.e. by colour change. The extent of enzymatic reaction is then a quantitative indication of the amount of antigen trapped by the primary antibody

Elution

Removal by dissolving, such as the removal of adsorbed material from an adsorbent by means of solvents

Enation

Laminar or cup-shaped outgrowth found on the underside of grapevine leaves affected by enation disease

Enzyme-labelled antibody

See Conjugated molecule and ELISA

Epinasty

An increase or decrease in growth of the upper or lower leaf surface or vein which causes the leaf to bend downward

Ethidium bromide

An intercalating agent which allows the ready detection of double-stranded nucleic acid molecules in agarose gels. The nucleic acid/ethidium bromide complex fluoresces brightly when exposed to ultraviolet (UV) light (ethidium bromide is highly carcinogenic)

Eye (of a bud)

The protruding meristematic portion of a bud which later enlarges and grows into a young shoot or flower

Flecking

As in leaf flecking. Usually a lighter translucent spot or small patch on leaves

Flush

The new, young and fresh growth of shoots and leaves

Foundation tree

In a certification programme, the foundation tree is the primary tree derived from budwood that has been specially selected, shoot-tip grafted and/or heat-treated. It has been indexed and certified as virus-free and also true-to-type. It will become the primary source tree for all future progeny trees. A foundation tree can be synonymous with a mother tree or mother block tree

Gel

The inert matrix used for electrophoretic separation of nucleic acids or proteins. Agarose gels are used for separation of DNA, agarose or polyacrylamide for RNA, and polyacrylamide for proteins

Graft transmission

The transmission of a virus or other pathogen(s) by grafting tissue from the suspect host to an indicator plant

Hybridization

The formation of stable duplexes between complementary nucleotide sequences via Watson-Crick base pairing. The efficiency of hybridization is a test of sequence similarity. DNA-DNA, DNA-RNA and RNA-RNA hybrids may be formed.

In classical genetics, and particularly plant breeding, hybridization means the formation of a novel diploid organism either by normal sexual processes or by protoplast fusion

Immunoassay

An assay system in which a protein is detected using an antibody specific to that protein. A positive result is seen as a precipitate of an antibody-protein complex. The antibody can be linked to a radioactive atom (radio immunoassay) or to an enzyme that catalyses an easily monitored reaction (see ELISA)

Immunoblotting

A procedure whereby either the antigen or antibody molecules are bound to a protein-binding substrate, such as cellulose nitrate, and then exposed to the complementary antigen or antibody. The antigen-antibody complex which

forms on the membrane is detected by an appropriately labelled antibody

Immunodiffusion

A procedure in which antibody and/or antigen molecules are allowed to migrate through an inert medium. A visible precipitate forms at the zone where related antigen and antibody molecules meet in a suitable concentration and react

Immunofluorescence

The effect when antigens are detected, often within tissues, by use of an antibody to which a fluorescent material is attached

Immunoglobulin

A blood serum protein that functions as an antibody, commonly a gammaglobulin

Immunosorbent

A material that can adsorb serologically active molecules (antigens or antibodies). Cellulose nitrate and some plastics, such as certain polystyrenes, are good immunosorbents. Adsorbed molecules typically retain serological functions

Inclusion bodies

Cytopathic intracellular structures found in virus-infected plants. They contain virus particles, other proteins or structures specific to the virus and/or formed as a result of virus infection

Indexing

Any testing that will consistently confirm the presence (or absence) of a transmissible pathogen or identify a disease. The index test should be specific for the pathogen or disease

Indicator plant

A plant used to test or index for the presence of a transmissible pathogen. The inoculated indicator plant will usually show very specific symptoms, thus permitting the diagnosis of a particular disease

Infection feeding
After acquisition feeding where the pathogen is ingested by the insect, infection feeding is the secondary or follow-up feeding where the pathogen is injected into the host plant by the insect

Inoculation
The process of infecting an indicator plant, usually by graft, mechanical or vector transmission

Inoculum tissue
Tissue containing the transmissible pathogen or pathogens

Intermediate antibody
An antibody used in the second step of an indirect ELISA assay. The intermediate antibody reacts to the antigen bound to the plate but is not labelled. It is detected by using another antibody which is labelled and is specific only for the intermediate antibody

Linearization
The conversion of a nucleic acid that is normally circular into a linear form of the molecule, done by cutting the circular form at a single site

Line pattern
Translucent, chlorotic or bright yellow lines, sometimes in a pattern resembling the outline of an oak leaf, induced in leaves of naturally infected vines or in indicators, mainly by nepoviruses

Loading
As in loading ELISA plates. In ELISA, it is the process of adding a given amount of sample, buffer or any substance to the wells of an ELISA plate or gel apparatus

Mealybugs
Scale insects of the family Pseudococcidae, vectors of some grapevine closteroviruses (see Leafroll)

Mechanical transmission
Transfer or transmission of graft-transmissible pathogens by means other than grafting and not involving vectors. This can be done by knife cut, razor slash, hand or cotton rubbing of sap on leaves using carborundum powder, or any other non-grafting method

Molecular hybridization
See Hybridization

Molecular probe
See Probe

Mollicutes
One of the four divisions of the kingdom Prokaryotae, characterized by having no cell wall or peptidoglycan. This division includes the mycoplasmas and mycoplasma-like organisms (MLOs)

Monoclonal antibody
An antibody preparation containing only a single type of antibody molecule. Monoclonal antibodies are produced naturally by myeloma cells. A myeloma is a tumour of the immune system. A clone of cells producing any single antibody may be prepared by fusing normal lymphocyte cells with myeloma cells to produce a hybridoma

Mother trees or mother block trees
Similar or synonymous with foundation or foundation block trees (see Foundation tree)

Negative stain

An electron-dense solution used to provide contrast around virus particles viewed on a transmission electron microscope

Nematodes

Small, soil-inhabiting, root-feeding, worm-like animals, vectors of several grapevine nepoviruses

Nick translation

A procedure to insert radioactive or other tagged bases in a DNA probe. It is a process whereby damaged dsDNA molecules are repaired with nucleotides, some of which are radioactive. It is a good way to repair a probe

Nitrocellulose membrane (cellulose nitrate)

A nitrated derivative of cellulose which is made into membrane filters of defined porosity, e.g. $0.45\,\mu m$, $0.22\,\mu m$. These filters have a variety of uses in molecular biology, particularly in nucleic acid hybridization experiments. In the Southern and Northern blotting procedures, DNA and RNA, respectively, are transferred from an agarose gel to a nitrocellulose filter. Some centrifuge tubes are made of nitrocellulose; they are readily punctured with a hypodermic needle, and are frequently used for sucrose gradient centrifugation

Northern blot, Northern transfer

A procedure analogous to Southern transfer, but in this case RNA, not DNA, is transferred or blotted from a gel to a suitable binding matrix such as a nitrocellulose sheet. Single-stranded RNA is separated according to size by electrophoresis through an agarose or polyacrylamide gel; the RNA is then blotted directly on to the support matrix with no denaturation. RNA fixed to the supporting matrix can then be hybridized with a radioactive single-stranded DNA or RNA probe

Nucleic acid

A DNA or RNA molecule, which can be single or double stranded

Nucleotide sequence

The order of alignment of nucleotides of nucleic acid molecules (see Sequencing)

PAGE

Polyacrylamide gel electrophoresis (see Polyacrylamide gels). A method for separating nucleic acid or protein molecules according to their molecular size. The molecules migrate through the inert gel matrix under the influence of an electric field. In the case of protein PAGE, detergents such as sodium dodecyl sulphate are often added to ensure that all molecules have uniform charge. Secondary structure can often lead to the anomalous migration of a molecule. It is common, therefore, to denature protein samples by boiling them prior to PAGE. In the case of nucleic acids, denaturing formamide, urea or methyl mercuric hydroxide is often incorporated into the gel itself, which may also be run at a high temperature. PAGE is used to separate the products of DNA-sequencing reactions, and the gels employed are highly denaturing since molecules differing in size by a single nucleotide must be resolved

Pegs

Needle-like growths or pinpoint projections observed on wood or bark, which may be symptomatic for certain diseases of grapevine. Pegs usually have corresponding pits on the opposite bark or wood surface

Peptidoglycan

Polysaccharide chains covalently cross-linked by peptide chains. The presence or absence of peptidoglycan in the cell walls of bacteria is used to distinguish gracilicute-like organisms (which contain peptidoglycans) from mycoplasma-like organisms (which have no peptidoglycans)

Plant laboratory

A sophisticated greenhouse designed and used primarily for indexing

Plate-trapped antigen

Antigen adsorbed directly on the ELISA plate without use of a trapping antibody. For example, virus particles can be trapped to the surface of ELISA plates from extracts of infected tissue added to the wells of the plate. Other proteins, however, are also adsorbed

Polyacrylamide gels

Often referred to, incorrectly, as acrylamide gels. Gels made by cross-linking acrylamide with N,N'-methylene-bis-acrylamide, used for the electrophoretic separation of proteins and also RNA molecules. DNA is usually too heavy to migrate far in polyacrylamide. Polyacrylamide beads are also used as molecular sieves in gel chromatography and are marketed under the brand name Nio-gel

Polyclonal antiserum

Antiserum harvested from the blood of immunized animals. Polyclonal antisera contain a mixture of antibodies to the various antigenic molecules present in the material used to immunize the animal. In contrast, monoclonal antisera contain only a single antibody

Potting mix

A mixture of ingredients used as an artificial soil medium for container growth of plants

Primary leaves

The first emerging leaves from a germinating seed. These leaves may be the cotyledons or may differ in shape from the secondary leaves

Primer

A low-molecular-weight species that promotes a reaction, such as an oligonucleotide that binds to a template, permitting a copy of the template to be further synthesized

Probe

Noun: a specific DNA or RNA sequence that has been radioactively labelled to a high specific activity. Probes are used to detect complementary sequences by hybridization techniques such as Southern or Northern blotting or colony hybridization

Verb: to hybridize in order to detect a specific gene or transcript, e.g. "We probed our bank with labelled viral RNA to detect clones containing viral DNA sequences"

Probe denaturation

Treating the probe under conditions that will separate its nucleic acid strands and permit their subsequent hybridization with the target molecules to be tested

Probe labelling

Investing probes with detectable tags by a variety of procedures

Prokaryotes

Bacteria-like organisms in the kingdom Prokaryotae that have no organized nucleus and are surrounded by a nuclear membrane

Protein A

A protein with a high affinity for antibody gammaglobulins

Radioactive probe

A nucleic acid that has been made radioactive by one of several techniques (e.g. nick translation) and is to be used to detect a complementary nucleic acid sequence

Recombinant plasmid

A bacterial plasmid DNA containing an insert of DNA from a non-related source, e.g. a plasmid containing an insert of viral cDNA. It is created by recombinant DNA technology

Replicative form (Rf)

The intracellular form of viral nucleic acid which is active in replication. For example, M13 phage particles contain a single-stranded DNA circle, while the Rf of the same molecule is double stranded

Restricted endonuclease

An enzyme that recognizes and cuts double-stranded DNA at specific sites determined by the sequence of bases at that site

Resuspension medium (RM)

See TKM buffer

Reverse transcription enzyme

The enzyme that accomplishes the enzyme synthesis of a copy DNA from an RNA template in the presence of a primer and nucleotide triphosphates under appropriate conditions

Ringspot

Circular, ring-like translucent, chlorotic or bright yellow spots induced in leaves of naturally infected vines or indicators, mainly by nepoviruses

RNA

Ribonucleic acid. The alternative reservoir of genetic information besides DNA. Viruses have single-stranded or double-stranded RNA genomes. In organisms, RNA is transcribed from DNA and is essential for the expression of the genetic information contained within the DNA. RNA differs from DNA in having ribose instead of deoxyribose as the sugar moiety in its nucleotides and in having uracil instead of thymine as one of its two pyrimidine bases. RNA, but not DNA, may be degraded by alkaline hydrolysis

Rugose

As in rugose wood, meaning rough or wrinkled, with pits and grooves on the woody cylinder

Secondary leaves

In contrast to the primary leaves of a germinating seed, the secondary leaves are permanent leaves with a fixed morphology. Many seedlings produce both primary and secondary leaves, but many have just one type

Sequencing

The determination of the order of nucleotides in a DNA or RNA molecule, or that of amino acids in a polypeptide chain

Sequencing gel

A long polyacrylamide slab gel that has sufficient resolving power to separate single-stranded fragments of DNA or RNA that differ in length by only a single nucleotide.

Electrophoresis is carried out at high voltage and with the gel in a vertical position. Urea is usually included in the gel mixture as a denaturing agent. This prevents internal base pairing within the single-stranded molecules and ensures that their relative speed of migration is solely dependent on their length

Shoot-tip grafting

A micro-grafting procedure. The meristematic growing tip (the meristem plus one to three leaf primordia) is excised by cutting the very young tip using the cutting edge of a razor-blade mounted in a special handle. This decapitated tip is then grafted to a very young seedling with the aid of a binocular microscope. The grafted plant is usually then grown *in vitro* and later transplanted or grafted to produce a shoot-tip grafted plant or tree

Shot berry

A disorder in which grape bunches bear small, imperfectly developed berries. It is induced and/or enhanced by viral infections (nepoviruses in particular)

Slot-blot

See Dot-blot

Southern blot, Southern transfer

A technique that combines the resolving power of agarose gel electrophoresis with the sensitivity of nucleic acid hybridization. DNA fragments separated in an agarose gel are denatured *in situ* and then blotted or transferred, usually by capillary action, from the gel to a nitrocellulose sheet or other binding matrix placed directly on top of the gel. Single-stranded DNA binds to the nitrocellulose and is then available for hybridization with labelled, ^{32}P or biotinylated, single-stranded DNA or RNA. The labelled nucleic acid is known as the probe and, in the case of DNA, is often prepared by nick translation. The hybrids are detected by autoradiography, in the case of ^{32}P, or by colour change, in the case of a biotinylated probe. A very sensitive and powerful technique, it is often described as "blotting"

STE buffer

Sodium chloride, Tris and EDTA (see Part III). Used in purification of nucleic acids including dsRNA and viroids

Stem pitting

Alteration of the xylem consisting of groove-like or pit-like depressions of the woody cylinder, as in rugose wood

Streptavidin

A microbial protein which binds biotin. It is preferred to avidin because of its more specific binding (see Biotin). In molecular hybridization, streptavidin reacts specifically with the biotin molecules fixed on the probe

Stub

As in a rootstock stub. The short projecting portion of the stem that remains after the rootstock is severed from the growing scion. The stub is that small portion above the scion which should be trimmed flush with the scion at a later time

Substrate

The substance acted upon by an enzyme. Usually this substance contains a chemical which, when acted upon by an enzyme, will undergo a colour change that can be easily seen and measured

Symptomless carrier

A tree or plant that contains a graft-transmissible pathogen but shows no symptoms. An example of a symptomless carrier would be an American rootstock containing leafroll-associated closterovirus

Synergism

A phenomenon in which the joint action of more than one agent such as two viruses in combination induces a more intense symptom in an indicator plant than the agents could induce separately

TAE buffer

Tris, sodium acetate·3H$_2$O and sodium EDTA

Template

The molecule that is acted upon or copied, as in the production of cDNA probes by reverse transcriptase

Template primer
See Primer and Template

Thermotherapy
Treatment of budwood or plants by heat to eliminate internal pathogens

TKM buffer
A buffer with Tris, KCl and MgCl$_2$

TME
As in TME Tris buffer. Tris, MgCl$_2$ and EDTA

Trapping antibodies
Antibodies used to coat the wells of ELISA plates (the first layer in sandwich assay procedures). Antibodies adsorbed to the solid surface of the plate trap related antigens from the sample extracts placed in the plate for testing

Triturating
Rubbing or grinding, as with tissue ground with pestle and mortar

Viroid
A small molecular RNA, transmissible in plants, which has no extracellular protein component or translation capacity and which can be pathogenic. It is composed of naked, single-stranded, low-molecular-weight RNA (MW 80 000 to 130 000) which utilizes only host components for its replication. Viroids exist in solution as rod-like structures arranged in a series of short base-paired and non-base-paired regions

Virus
Viruses are macromolecular transmissible agents capable of causing diseases in plants and animals. They are small enough to pass through a millipore filter of 0.2 μm. They have been considered to be either living organisms or simply a molecular complex of nucleic acids and proteins capable of multiplication in living cells. Viruses are characterized by a core of nucleic acids with a genome of less than 3 x 10^8 daltons surrounded by a protein coat that can induce formation of antibodies

Western blot
A procedure used to determine the presence of viral coat protein in plant tissue extracts. Extracts are electrophoresed in polyacrylamide gel slabs; protein bands are then electrotransferred from slabs to an appropriate support matrix (e.g. PVDF membrane) and exposed to specific antiserum for recognition

Yellows
Disease induced by mycoplasma-like organisms (MLOs), characterized by rolling of leaves, generalized or sectorial yellowing or reddening of the leaf blade, irregular ripening of canes and withering of bunches